# Israel's Way of War

# Israel's Way of War

## *A Strategic and Operational Analysis, 1948–2014*

EHUD EILAM

McFarland & Company, Inc., Publishers

*Jefferson, North Carolina*

LIBRARY OF CONGRESS CATALOGUING-IN-PUBLICATION DATA [new form]

Names: Eilam, Ehud, author.
Title: Israel's way of war : a strategic and operational analysis, 1948–
    2014 / Ehud Eilam.
Description: Jefferson, North Carolina : McFarland & Company, Inc.,
    Publishers, 2016 | Includes bibliographical references and index.
Identifiers: LCCN 2015046596 | ISBN 9781476663821 (softcover :
    alkaline paper)
Subjects: LCSH: Military art and science—Israel. | Israel—History,
    Military—20th century. | Israel—History, Military—21st century. |
    Low-intensity conflicts (Military science)—Middle-East—History. |
    Israel—Strategic aspects. | Operational art (Military science)
Classification: LCC UA853.I8 E55 2016 | DDC 355.4095694—dc23
LC record available at http://lccn.loc.gov/2015046596

BRITISH LIBRARY CATALOGUING DATA ARE AVAILABLE

ISBN (print) 978-1-4766-6382-1
ISBN (ebook) 978-1-4766-2325-2

Front cover: Israeli infantry atop armored personnel
carriers while others walk behind them (photographer Avi
Ohayon; Israel Government Press Office)

Printed in the United States of America

*McFarland & Company, Inc., Publishers*
    *Box 611, Jefferson, North Carolina 28640*
        *www.mcfarlandpub.com*

# Table of Contents

# Preface

The Arab-Israeli conflict has been the subject of numerous books, dealing with many of its aspects. This volume focuses on the military facet of this ongoing confrontation. Its aim is to explore various perspectives of Israel's high-intensity, hybrid and low-intensity wars, such as the linkage between high- and low-intensity wars as reflected on strategic and operational levels.

Israel and other nations, such as the United States and Britain, have to be prepared at all times to conduct high-intensity, hybrid or low-intensity wars. It is therefore important to understand how the IDF (Israeli Defense Forces) prepared and conducted the various conflicts in which it has been involved, and to consider the unique challenges of each.

This work is more of a personal treatise, reflecting my own analysis and insights, presented in a descriptive style.

Nevertheless, it includes hundreds of notes supporting my ideas. This book is based on various sources: books, articles, and so on, including dozens of documents from IDF's archives from the 1950s to the early 1970s. All of the sources used are unclassified.

Following my three years of regular service in the IDF, where I did field duty in the anti-aircraft corps, I focused on studying Israel's national strategy and military doctrine. I have been involved academically and practically in this field for more than twenty years, and wrote my MA and PhD theses on these matters. I also worked for a few years for Israel's Ministry of Defense as a private contractor in my line of expertise. This book is a completely personal project, and not part of any research I did for the Ministry of Defense. This work expresses my personal views, and does not necessarily represent the opinions of others. I began working on it in Israel and have finished it in the U.S., where I now reside.

I wish to thank my parents and my two elder brothers for their warm and solid support over the years. I would also like to thank Dr. Gil Ribak for his advice. Of course, all the errors here are mine alone. For readers with questions or comments about the book, please write to Ehud Eilam at Ehudei2014@gmail.com.

# Introduction

Since its establishment in 1948, Israel has had to handle a variety of high-intensity, hybrid and low-intensity wars. This book examines how Israel has dealt with those diverse confrontations by evaluating political, strategic and (primarily) military factors, such as combat patterns of the IDF.

From the beginning, Israel had several basic strategic challenges that had to do with high-intensity wars, such as lack of strategic depth and asymmetry between it and the Arabs, which required an adequate combat doctrine. At the same time, Israel had to deal with low-intensity wars, which had a connection to high-intensity wars (such as in the way they were conducted).

Hybrid warfare can be defined as the "employment of the combination of traditional, irregular, catastrophic, and disruptive tactics, techniques, and procedures in an effort to achieve success, across the full range of warfare: tactical, operational, and strategic."[1] "Tactics including ambushes, terrorism, improvisation, information warfare and other forms of asymmetric, unconventional warfare characterize" hybrid outfits.[2]

Since the early 1980s, Israel has dealt with hybrid foes in four wars: the PLO in June–September 1982, the Hezbollah in July–August 2006, and the Hamas in December 2008–January 2009 and July–August 2014. Those confrontations are analyzed in this book according to their theater of operations: the 1982 and 2006 wars in Lebanon and the 1956, 1967, 2008–2009 and 2014 clashes in the Gaza Strip.

Another aspect of this book is the similarity between Israel and Western states (mostly the United States and Britain) in fighting high- and low-intensity wars. Special attention is devoted to comparing the Vietnam War and Israel's war against Hezbollah in Lebanon in the 1990s.

Israel's high-intensity wars were its interstate confrontations with

Arab states (1948–1949, 1956, 1967, 1973 and 1982). Those were, for that region, huge conventional collisions that took place on the ground, air and sea. They usually lasted a few days or weeks. Those campaigns involved tens, sometimes hundreds of thousands of troops, along with hundreds or thousands of various military platforms. Israel conquered vast areas and destroyed large Arab forces as a result of these wars.

Most of Israel's low-intensity wars were border wars, taking place between Israel and Arab states and/or organizations. Arab troops/fighters and Israeli soldiers took part in those struggles, usually involving anywhere from a few men to hundreds. The brawls were often short—hours, if not minutes—while the confrontations themselves went on for years. The hostile actions across the borders were quite diverse: laying down ambushes and explosive devices, combat encounters (some during pursuits), infiltrations for all kinds of purposes, raids and exchange of fire ranging from small arms to tanks, artillery and planes. Israel also went through low-intensity wars that were not around its borders, but rather within them, in 1987–1993 and 2000–2005.

Hybrid wars "combine the strengths of an irregular fighting force with various capabilities of an advanced state military, and will play an increasingly prominent role in international security issues."[3] For Israel, a hybrid war was a combination of low- and high-intensity wars, including those that occurred during 1982 and 2006 in Lebanon and during 2008–2009 and 2014 in the Gaza Strip. Those confrontations lasted about a month or two. Israel's foes (the PLO, Hezbollah and Hamas) were a mixture of guerrilla and terror organizations with conventional militaries, preparing and fighting accordingly. The campaigns were not just a series of short clashes and exchange of punches, like in Israel's low-intensity wars, nor were they only based on a massive and rapid Israeli offensive, as in Israel's high-intensity wars. Israel's hybrid wars were instead a kind of mix of those two types of combat.

The first chapter discusses Israel's strategic and political constraints, connected to its high-intensity wars. After the 1948–1949 war, Israel created its national security policy, which contained some fundamental principles. A defeat in a high-intensity war would have jeopardized the existence of the new state. The balance of power, including population size, land and natural resources, was totally in favor of the Arabs. Israel also lacked strategic depth in some of its fronts until 1967.

Although the IDF won high-intensity wars, Israel could not destroy the Arabs' military capability or force Arab governments to accept Israel's basic right to exist—not even after the decisive victory of 1967. Israel also could not hope to wear down its foes, since the Arabs (particularly states such as Egypt) could refill their ranks again and again thanks to their large populations. The same phenomenon has happened with weapon systems. The Soviet Union, when it was the patron of Syria and Egypt, sent new platforms to replace those that were lost in combat. It seems, therefore, that the IDF's work is Sisyphean, and all it annihilated or captured in battle from the Arabs has been to a large extent in vain.

During high-intensity wars against Israel, Arab states strove to maximize their potential by forming military coalitions. However, in the 1956 and 1967 wars this alliance was symbolic. Only in 1973 did the Arab states—mainly Egypt and Syria—manage to confront Israel together in an effective way. Israel, for its part, had to rely on its own forces, except in 1956, when Israel collaborated with France and Britain against Egypt— that is, Gamal Abdel Nasser; the two European powers played an important role in the campaign between Israel and Egypt by destroying the Egyptian air force. Israel also developed its relationship with the United States, which did not participate in Israel's high-intensity wars but assisted it militarily and politically.

The second chapter analyzes Israel's combat doctrine in high-intensity wars, strategically based on preventive war or preemptive strikes. The enormous buildup of the Egyptian military in 1956, following its huge arms deal with the Soviet Union, was the main reason for an Israeli preventive war against Egypt. In 1967 Israel took the initiative once more when its forces launched a preemptive strike. IDF's campaign tactics in a high-intensity war relied on attack, striving for a rapid victory while destroying Arab forces and conquering key areas. In both the 1956 and 1967 wars, the IDF seized the Sinai and the Gaza Strip in less than a week, while at the same time annihilating or pushing several Egyptian divisions all the way to the Suez Canal.

Achieving air superiority was vital. Throughout the years the IAF (Israeli Air Force) has been engaged in a long struggle with Arab jets and anti-aircraft batteries, particularly in 1973. During that showdown, Israel's armor troops demonstrated their various skills, engaging in tank-on-tank battles and so forth, but Israel's tanks were often vulnerable (and sometimes helpless) against Egyptian anti-tank fire. If there was an Israeli corps that excelled in the 1973 collision, it was the navy. This branch is responsible for guarding the coasts, where most of the Israeli population is located, and securing routes in the Mediterranean Sea and the Red Sea. In 1973 Egypt and Syria tried to block Israel's sea-lanes but failed to do so, and their navies experienced a clear defeat in the main arena—the Mediterranean.

The third chapter describes the linkage between high- and low-intensity wars, expressed in different ways For example, Israel suffered from infiltrations of Palestinians in the early 1950s. In those low-intensity wars—basically border wars—Palestinians exploited the short distance from the border to Israeli towns, just as Arab militaries would have done in a high-intensity war.

In the border wars of the 1950s Syria, as well as Egypt and Jordan, clashed with Israel directly, but each of those Arab states essentially fought alone. This pattern repeated itself in high-intensity wars in 1956 and 1967. Egypt was attacked by Israel, but Syria and Jordan did not really assist their Arab ally. In 1967, while Egypt and Israel collided in a high-intensity war, Syria and Jordan ran a border war against Israel.

Sometimes a low-intensity war could ignite a high-intensity war, as did the struggle between Israel and Syria over demilitarized zones and water sources in the 1960s. A low-intensity war on one front could also cause a high-intensity war on another front. In February–March 1960 Egypt sent considerable reinforcements into Sinai after an Israeli raid on a Syrian post. Egypt wished to deter Israel from attacking Syria again, which brought about a crisis that was eventually resolved peacefully, but it could have caused a high-intensity war between Israel and Egypt.

The 1956 and 1967 confrontations proved that in spite of successful strikes in the previous border wars, a full-scale offensive in a high-intensity war was required. The 1956 war stopped the infiltrations from the Gaza Strip, and the 1967 showdown ended border disputes on the Syrian front.

Inversely, high-intensity wars sometimes caused low-intensity wars. Such was the case in the 1967 showdown, rolling low-intensity wars up to the Egyptian, Jordanian and Syrian fronts. In those collisions, Israel confronted both Arab states and guerrilla and terror organizations.

A defeat of the IDF in a high-intensity war could have cost Israel its independence. In this scenario, some of the Israelis may well have become insurgents conducting a low-intensity war against Arab rule of their land. If Arab states had lost a high-intensity war, their regimes might have faced a low-intensity conflict—that is, a rebellion in their own countries. Another scenario of an internal brawl inside an Arab state could have developed following a low-intensity war with Israel. (This actually happened in 1970 in Jordan after the PLO used the Hashemite kingdom to collide with Israel.)

The linkage between high- and low-intensity wars in the Middle East also relates to foreign powers. Britain, which had strong ties with Jordan in the 1950s, was almost entangled in a high-intensity war against Israel because of the border wars between the latter and the Hashemite kingdom. However, when Britain eventually became involved in a high-intensity war in that region, it was together with Israel and France against Egypt. One reason why France participated in this high-intensity war was to speed up the end of its own low-intensity war in Algeria by bringing down Nasser's regime, which supported the Algerian mutineers. As to the superpowers, the Soviet-Egyptian arms deal from September 1955 caused Israel to rearm. The Egyptian and Israeli estimate regarding the time their enemies would need in order to be ready for a high-intensity war influenced their strategies in the ongoing border wars.

Arab states wished not to be entangled in the 1987–1993 and 2000–2005 low-intensity wars between Israel and the Palestinians, in order to avoid a high-intensity war with Israel. Egypt in particular was careful not to be dragged into the fight in the Gaza Strip. This approach continued after the 2011 revolution.

The fourth chapter deals with the way the IDF has conducted high-intensity, hybrid and low-intensity wars. The IDF was established at the beginning of the 1948–1949 war, in late June 1948, based on several armed organizations, active in the underground before the establishment of the state in 1948. The IDF fought a high-intensity war, but its troops were in poor condition (at least in the first stages of the showdown) because of lack of training, weapons, and so forth. This explains the Arab perspective of the confrontation as being a hybrid/low-intensity war, since their militaries were more modern and organized.

In the 1950s Israel suffered from infiltrations. The IDF preferred not to handle this sort of low-intensity conflict, since it was at the expense of preparing for a high-intensity war, its most important mission. Yet the IDF could not have avoided those border wars, since the military was usually the only entity capable of dealing with this challenge.

There was a consistency in the IDF's combat patterns in both high- and low-intensity wars. The IDF relied on taking the initiative, striving for a rapid victory and destroying Arab forces, while at the same time absorbing minimum casualties. The IDF also used methods of low-intensity warfare, such as raids, in high-intensity wars. Yet early warning, a vital component of its doctrine, was much more critical in a high-intensity war because of the ramifications of a full-scale Arab offensive, compared to an Arab incursion in a low-intensity war.

Until 1956, the buildup of the IDF was based on the infantry, given the successes of this corps in the 1948–1949 war and the lack of awareness regarding the potential of the armor corps. Embarrassing failures of the infantry in the border wars raised doubts about its future capability in a high-intensity war. But only after the 1956 war did the IDF become convinced that it should depend on the armor and air force instead of the infantry. The accomplishments of the armor and the air force in the border wars of the 1960s, particularly in the 1967 high-intensity war, reinforced their status. It also explains why the IDF continued to trust those two corps in the border wars in the years 1967–1973, as well as during the high-intensity war of 1973.

On the tactical and operational levels, the experience the IDF gained in border wars helped it prepare for similar tasks in high-intensity wars, such as examining troops under fire, conquering posts in the 1950s and confronting Syrian tanks and jets in the 1960s.

In the border wars in the 1950s, the "101" elite unit was supposed to be a role model for the rest of the IDF, in low-intensity wars as well as high-intensity ones. The 1956 war proved that the performance of some units was inadequate (such as that of the 10th Brigade in Um Katef). Furthermore, many aspects of high-intensity war, such as command and control of a division, or mass air-to-ground bombardments, could not have been tested in border wars at all. In other cases, being used to border wars was a drawback during a high-intensity war. In the early 1970s armor units in the Golan were accustomed to "combat days" (limited clashes that occurred only once in a while). At the beginning of the 1973 showdown, Israeli tank crews and their commanders believed they were facing another familiar collision; instead, they found themselves in the midst of a full-scale Syrian offensive. The border war turned out to be a high-intensity war.

Starting in the mid-1980s the IDF had to transfer a large part of its attention from preparing for another high-intensity war—which never came—to dealing with low-intensity and hybrid wars. Training for and confronting low-intensity wars (1987–1993 and 2000–2005), together with skirmishes in the 1990s with a hybrid foe—the Hezbollah—kept the IDF very busy. Furthermore, menacing long-range missiles and rockets replaced Arab bombers as the most dangerous threats to the Israeli rear. The same could be said about suicide bombers who infiltrated Israeli cities, as Arab tanks might have done in a high-intensity war.

The fifth chapter examines the Gaza Strip as a battlefield in four clashes: between Israel and Egypt in 1956 and 1967, and between Israel and the Hamas in 2008–2009 and 2014. Two of the collisions were part of a high-intensity war, in 1956 and 1967, while in 2008–2009 and 2014 it was a hybrid war. Israel attacked the Gaza Strip in 1956, 1967, 2008–

2009 and 2014, since that area served or could have served as a jumping-off point against the south of Israel. This threat was realized over the years by means of both infiltrations and firing on Israeli targets (mostly civilian ones). As a result, the IDF attacked while exploiting the long and narrow terrain of the Gaza Strip in order to penetrate quickly and deeply into that area. In 1956, 1967 and 2008–2009 this military feat was made even easier by the fact that the defender of the Gaza Strip did not put up much of a fight.

The sixth chapter compares two of Israel's wars against hybrid foes: the PLO in 1982 and the Hezbollah in 2006. The 1982 and 2006 confrontations were the peak of the conflict between Israel and those two non-state organizations that had turned Lebanon into a base of operations against Israel. Strategically, Israel failed in both wars to achieve major goals such as destroying its enemy. Israel experienced more success in 1982, when most of the PLO was forced to leave Lebanon, but it took more than two months of combat. In both wars the IDF faced similar challenges, such as urban warfare, avoiding anti-tank ambushes, and so forth. The IDF was also better prepared for the fight in Lebanon in 1982, as compared to 2006.

The seventh chapter discusses the strategic and military linkage between Israel and Western states in handling high- and low-intensity wars. Common strategic issues included the threat to Israel's very existence as a state during a high-intensity war, establishing the credibility of allies, surviving ongoing struggles, facing a coalition and forcing regime change. Militarily, the IDF implemented patterns of Western combat doctrine related to high-intensity wars such as implementing a forward defense, launching deep armor penetrations, relying on air power, and so forth. As to low-intensity wars, Israel had military and moral concerns similar to those of the United States in Vietnam and France in Algeria.

The eighth chapter compares the Vietnam War and the Israeli-Hezbollah war in the 1990s as a kind of case study, describing the way Israel and the United States conducted their respective low-intensity wars. In spite of the major differences between the two conflicts, the United States and Israel had to deal with an enemy that had a strong patron (the Soviet Union in Vietnam, and Syria and Iran in Lebanon). The United States and Israel had important advantages in technology, firepower, weapon systems, well-trained troops, and so forth. Yet they also had political constraints and military problems, such as the terrain, that gave plenty of cover to their foes. Adjusting to the guerrilla and terror methods of their rivals took time, while criticism at home intensified. Winning was therefore not easy, particularly since the local allies of the United States in Vietnam, and those of Israel in Lebanon, were weak and required strong support in order to survive.

The ninth chapter explores the war in Libya in 2011, a mutiny occurring because of internal reasons, but an event that could have happened since 1948, stemming from the potential defeat of an Arab regime in a war against Israel. At a certain stage, other states (mostly Western powers) joined the battle against the Libyan government. This confrontation was an example of a low-intensity war between Arabs, with no Israeli involvement, and high-intensity warfare in the Middle East from a Western point of view. The rebels and Western powers had several constraints. The Western powers faced the dilemma of whether to intervene, based on the evaluation of which of the options jeopardized the people of Libya and the security of Europe more. The challenge of how to assist the poorly armed rebels, many of whom lacked any kind of military training, followed.

All in all, this book examines the manner in which Israel fought its wars in the years 1948–2014. The aim is to provide different perspectives on the long, complicated and diverse conflict between Israel and the Arabs.

# 1

# Israel's National Strategy
# and Its High-Intensity Wars

Following the 1948–1949 war, David Ben-Gurion, Israel's first prime minister and defense minister, declared Israel's wish for peace.[1] Israel and Arab states such as Egypt, Jordan and Syria signed armistice agreements, although the negotiations did not lead to peace accords.[2] Over the years there were all kinds of official and non-official understandings between the two sides, regulating their relationship. For example, Israel and Egypt reached a binding peace treaty only in 1979.

Israel's national defense policy was created to a large extent following the 1948–1949 war.[3] Although it was never declared as the official Israeli policy, Israel accepted its principles, which remained more or less unchanged over the years.

## Winning the War

On 15 May 1948, when Israel was established, it was attacked by several Arab militaries that aimed to destroy it. Ever since that war, which Israel refers to as the War of Independence, the leaders of the new state have been extremely concerned about the severe ramifications of a major strategic defeat in a high-intensity war. Israel, mostly at the beginning of the 1948–1949 and 1973 wars, was in fact beaten in some battles and was more than once close to losing the entire showdown.

Moshe Dayan, Israel's minister of defense in the 1973 war, believed that Israel had defeated its enemies in that conflict.[4] But Arial Sharon, a division commander in the same war, claimed in a closed session in 1974 that "strategically, we did not win."[5] On 22 October 1973, Henry Kissinger, the U.S. secretary of state, told the Israeli prime minister that Israel "won the war, though at a very high cost," and Egypt "achieved nothing."[6] Yet Egypt and Syria declared that they had achieved victory in the 1973 war.[7] Either way, that showdown was Israel's ultimate test since 1948. On the one hand, the war proved that the Arabs' best efforts could not have subdued Israel. On the other hand, in the coming years Arab militaries, which showed improved performance when compared with former clashes, might have been upgraded even more, and to such a degree that they could have reached a decisive victory in a high-intensity war.

## *The Balance of Power*

Israel has long been very much aware of the enormous advantages the Arabs possess due to their superiority in key strategic factors, such as the magnitude of their population, territories and natural resources, which obviously increased the probability of an Arab triumph in a high-intensity war. As a response to the Arab superiority in numbers, Jewish immigration to Israel was encouraged, while at the same time supporting local population growth, although there was no chance of catching up with high Arab birth rate. Israel therefore had to focus on upgrading the quality of its manpower as much as possible. In spite of those initiatives, Israel understood that it could not overcome the Arabs. Israel beat Arab states again and again while conquering vast areas, as was the case in 1967, but it was insufficient to force Arabs to recognize Israel's right to exist. Indeed, one of the major questions concerning the Arab-Israeli conflict has been whether Israel's achievements in high-intensity wars will ever convince Arabs to reconcile. There could, of course, be another outcome to those wars: Israel could eventually lose them, or crumble from the inside because of social and economic troubles and the ongoing Arab pressure.[8] Nevertheless, Israel proved that it has a vibrant society, and it has become in recent years "a global leader in microchip design, network algorithms, medical instruments, water recycling and desalinization, missile defense, robotic warfare, and unmanned aerial vehicles."[9]

Following the Israeli-Egyptian peace treaty, Egypt (at least publicly) has ceased to consider Israel its enemy. This substantial change did not create tens of millions of Egyptians supporting Israel, but they were still taken out of the Arab-Israeli strategic equation, at least in terms of working against Israel in a high-intensity war. Israel and Egypt had a "cold peace," and sometimes it looked like a "cold war," but even the worst crisis between them did not initiate a high-intensity war. This situation has lasted for more than 30 years, which seems forever in the fragile, dubious and too often irrational Middle East.

Syria and Israel continued to be foes preparing for a war that never happened. Syria—becoming a "powerful regional actor in Middle Eastern politics"[10] —has tried since the early 1980s to take the place of Egypt as the head of the anti-Israel camp. The main reason driving this policy was Syria's desire to use the struggle against Israel in order to obtain hegemony in the Arab world. In terms of population size and land alone, Syria—like Egypt—is much bigger than Israel. However, Syria is much smaller than Egypt in terms of both people and territory. Furthermore, Syria, becoming independent in 1946, does not possess the historical status of Egypt. And Hafez al Assad, Syria's leader in 1970–2000, was not Gamal Abdel Nasser, the legendary and charismatic Egyptian ruler who intimidated his rivals but earned enormous popularity among his people. Assad, who was much less exciting, could only wish to possess the influence and position Nasser had enjoyed.

Iraq, under the leadership of Saddam Hussein, was a known opponent of Assad and as hostile toward him as it was toward Israel, if not more. The alliance between Iran and Syria did little to gain friends for the latter in the region, not only from the Iraqi perspective during its brutal showdown with Iran but also among other Arab states fearing the huge Shiite/Persian state. In the 1980s, when Iran faced off against most of the Arab world, Assad found it difficult to gather other Arabs against the non-Arab state in the region, Israel. When his attempt to create a united front against Israel failed, Assad strove to confront Israel on his own. This was his "strategic parity" vision. The collapse in 1991 of Syria's patron, the Soviet Union, was a major blow to Assad's ambitions.

In 2013 political frustration and economic hardships in Arab states might have caused them to blame Israel as an outlet.[11] Yet Efraim Inbar, director of the Begin-Sadat Center for Strategic Studies, said in May 2013 that "the most significant result of the Arab upheavals in recent years is the weakening of the Arab state, which has increased the power differential between Israel and its neighbors."[12] Indeed, there was a rapid decline in the strength of Iraq and Syria, two of Israel's more powerful foes. Furthermore, the peace treaties between Israel and both Egypt and Jordan have survived. The basic asymmetry between Israel and Arab states did not change, but Israel was in a relatively convenient strategic situation, particularly regarding the low probability of another high-intensity war.

AFGHANISTAN
Kandahār
Herāt
PAKISTAN
Karachi
TURKMENISTAN
Aşgabad
Masthad
ISLAMIC REPUBLIC OF IRAN
Tehrān
Eşfahān
Shīrāz
Tabrīz
Caspian Sea
Arabian Sea
Gulf of Oman
Muscat
OMAN
Dubayy
Abu Dhabi
UNITED ARAB EMIRATES
QATAR
Doha
BAHRAIN
Al Manāmah
Ad Dammām
Al Hufūf
Riyadh
SAUDI ARABIA
Kuwait
KUWAIT
Persian Gulf
Al Başrah
Baghdād
IRAQ
Arbīl
Kirkūk
Al Mawşil
TURKEY
Halab
SYRIAN ARAB REP.
Damascus
Tripoli
Beirut
LEBANON
ISRAEL
Port Said
Suez
CYPRUS
Mediterranean Sea
Alexandria
Cairo
Al Minyā
Asyūt
EGYPT
Aswān
Ammān
JORDAN
Al 'Aqabah
Medina
Mecca
Jeddah
Red Sea
Khartoum
SUDAN
SOUTH SUDAN*
LIBYA
Jīzān
ERITREA
Asmara
ETHIOPIA
YEMEN
Sanaa (Şan'ā')
Al Hudaydah
Al Mukallā
Gulf of Aden
Aden
DJIBOUTI
Djiboud
SOMALIA
Şuqutrā (Socotra)

The boundaries and names shown and the designations used on this map do not imply official endorsement or acceptance by the United Nations.

*Final boundary between the Republic of the Sudan and the Republic of South Sudan has not yet been determined.

800 km
500 mi

Department of Field Support
Cartographic Section

Map No. 4102 Rev. 5 UNITED NATIONS
November 2011

## Strategic Depth

Since the establishment of the State of Israel, the Tel Aviv district has been a critical area,[13] being both in the middle of the country and a center of population, industry, transportation, and so forth. The Haifa area, which is about 100 kilometers north of Tel Aviv, has been another vital zone because of its relatively large population, naval port and industry (such as refineries). The Jerusalem area, of course, has its own unique importance because of its vast population and historical, religious and national status.

As far as maintaining the war effort in a high-intensity war—using the military and civilian industries, infrastructure and transportation—the Haifa and the Tel Aviv districts have been the most important ones. This decisive factor was clear to Israelis and Arabs alike.

About 25 kilometers east of the Tel Aviv area and about 30 kilometers southeast of the Haifa zone there is the West Bank, which was under Jordanian rule in the years 1948–1967. Israel could have faced on that front not only the Jordanian military, which has been relatively small, but also units from other Arab militaries reinforcing the Hashemite kingdom. In the past, it was not easy for Israel to spot quick military movements into Jordan from remote Arab states such as Iraq in time to prepare for such attacks. However, it was a complicated task for the Arabs to coordinate this kind of maneuver, due to the political and military constraints involved in gathering a force strong enough to beat Israel.

Until 1967, the West Bank served as a kind of huge Arab bulge inside Israel, a comfortable position for Arab militaries, which could have turned not only toward the Tel Aviv or Haifa areas but also to other districts, thus threatening the flanks and rear of Israeli units. A sharp Arab strike from the West Bank could have cut Israel in half in a few hours. This would have made it almost impossible for the IDF to send troops by land from north to south, or vice versa, especially armor, since the Israeli navy was never built to carry large amounts of armored vehicles such as tanks from shore to shore.

The Golan—that is, the Golan Heights—is about 180 kilometers northeast of the Tel Aviv area and about 85 kilometers east of the Haifa area. In the years 1949–1967 the Golan was the border between Israel and Syria. Until 1967 Israeli villages, such as the kibbutz and other places in the Jordan valley near the Syrian border, were often under fire, mostly in the 1960s. In 1967 Israel seized the Golan and added about 20 kilometers to its strategic depth. This was more meaningful for the Haifa area than for Tel Aviv, which was quite far from the Golan. Having the Golan in Israeli hands was, for towns and villages in the Jordan valley, a huge relief. Yet the establishment of settlements in the Golan exposed Israeli civilians once again to Syrian aggression, as was demonstrated in the 1973 showdown. This threat never went away. Nevertheless, the Golan was essential to Israel in case of another massive Syrian attack.[14] Syria justified its offensive in 1973, claiming it was merely an attempt to repossess its lost land. Furthermore, although the deployment of the IDF in the Golan, about 65 kilometers from Damascus, has been serving as a deterrent since 1967, it was also a constant provocation, which encouraged Syria to try and retake the Golan.

Over the years Syria acquired long-range surface-to-surface missiles[15] that bypassed the Golan and covered all of Israel. Syria's missiles could not have inflicted many casualties unless they had carried chemical warheads, but in that case the Israeli response would have been devastating to both Syria and the Assad regime—perhaps even to Assad personally. Overall the latter avoided starting a direct offensive against Israel. Assad wanted the Golan only as long as this endeavor was not too costly.

Israel and Syria have had several rounds of peace talks in the past, the last one in 2008. Syria wished to receive the Golan,[16] but Israel had its concerns about that idea. Even with the Golan as a demilitarized zone, Syrian troops could launch a surprise attack. Syrian airborne commandos could also hold key sites in the Golan until the Syrian armor joined them.[17] This is why Israel wanted to ensure that the Syrian military would be moved as far as possible from the Golan. In 1967 and the 1973 showdowns Israel tried to reach a similar result: pushing back the Syrian military. In 1967 Israel did so by storming the Golan, whereas in 1973 it attacked from the Golan itself. By means of either peace or high-intensity wars, Israel strove to extend its strategic depth. (The same approach was used on the Egyptian front as well, when Israel kicked the Egyptian military out of Sinai in the 1956 and 1967 wars or kept its troops away from most of the peninsula following the peace treaty of 1979).

In 2008 the break in negotiations between Israel and Syria could have caused a war.[18] The latter could have decided to gamble once again on a high-intensity war, as in 1973—that is, trying to retake all the Golan by force. Yet, considering the fact that Syria was bothered by military problems starting in the early 1990s, it made more sense for Syria to implement only a limited high-intensity war aimed at seizing part of the Golan, while hoping for international pressure on Israel to return the rest of that area.

Syria could have used Lebanon to attack Israel.[19] Lebanon is about 200 kilometers north of the Tel Aviv area and about 100 kilometers north of the Haifa area. Ever since the late 1970s, when the Syrian army entered Lebanon, Syrian units were able to exploit Lebanon as a corridor into the north of Israel, including attacks aimed at the Golan. Yet its deployment there was tens of kilometers from the Israeli border with Lebanon, which added to Israel's considerable strategic depth on this front. The Syrian presence in Lebanon was also quite limited in comparison to Syrian deployment in the Golan, and therefore it did not cause much concern for Israel. Actually, Syria was more worried about its own strategic depth, since the IDF could have invaded Lebanon and put at risk the southwest part of Syria, its center of gravity. For this reason alone, the 1982 confrontation between Israel and Syria in Lebanon could have escalated into a full-blown war. For Syria the Israeli proximity to Damascus, in the Golan, was dangerous enough. A strong Israeli grip in Lebanon, particularly near the Syrian border, would have been a major worry for Assad.

In 2005 Syria was forced to pull its troops out of Lebanon.[20] This withdrawal shortened Syrian military communication lines on the Lebanese front and concentrated its troops in the Golan. However, considering the weakness of the Syrian military since the 1990s, a massive offensive against Israel was unlikely. The civil war that started in Syria in 2011 has to a large extent occupied and fragmented the Syrian military, thus cancelling any option of a Syrian attack to regain the Golan.

The Gaza Strip, which was under Egypt's control in the years 1949–1967, is about 80 kilometers south of the Tel Aviv area. Considering the strength of Egypt's military, the value of the Gaza Strip as a springboard for a surprise attack in a high-intensity war was alarmingly clear. In 1949–1967 Egypt could have also cut off the south of Israel, the Negev desert, from the rest of the country by advancing from the Gaza Strip toward the West Bank (Jordanian territory), a distance of about 40 kilometers.

In the years 1949–1967 Egypt could have potentially invaded Israel from the Sinai Peninsula, which is at least 150 kilometers south of the Tel Aviv area. Egypt usually deployed there only a small percentage of its troops. Yet the IDF found it difficult to detect in advance a large and sudden transfer of Egyptian forces from deep inside Egypt to the Israeli border. The conquest of the Sinai in 1967 gave Israel more strategic depth.

The peninsula was returned to Egypt in the early 1980s following the peace treaty between the two states. Since then, Israel has basically returned to its concept from the years 1949–1967: as long as the peninsula is mostly demilitarized, the danger to Israel is relatively minor. Since the early 1980s Israel has also had an Egyptian official obligation to keep the peace, which coincided with the Egyptian policy of the years 1949–1967, based on Egyptian awareness that dispatching troops into Sinai will risk a high-intensity war with Israel.

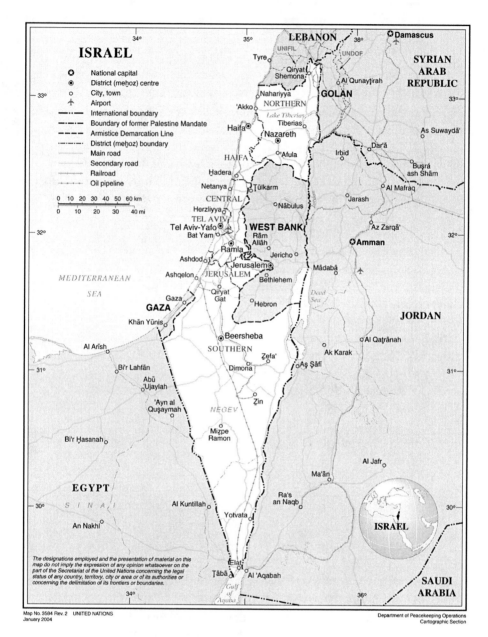

Map No. 3584 Rev. 2    UNITED NATIONS
January 2004

Department of Peacekeeping Operations
Cartographic Section

## *Arab Military Coalitions*

One of shortcomings of the Arabs in the 1948–1949 war was the lack of mutual cooperation between them.[21] The Arabs could have possibly overcome this difficulty by putting aside their internal rivalries, if their ambition to confront Israel had been a top priority. Another Arab problem in the 1948–1949 war was the fact that their states were relatively new entities and their militaries were not experienced in a high-intensity war, particularly as part of a coalition.

In the 1956 war the IDF concentrated 10 of its 16 brigades in Sinai, leaving only six brigades in other sectors, facing Syria and Jordan.[22] Those two states were allied with Egypt,[23] but they did not assist Egypt, which was left on its own. This saved the IDF from having to allocate more of its units to the Syrian and Jordanian fronts, thus making the 1956 war quite unique in the Arab-Israeli conflict, since Israel did not really deal with an Arab coalition.

In May 1967 Nasser might have been aware that the IDF had the edge.[24] Yet the Egyptian leader was convinced that Israel was getting ready to attack Syria, and in response Egypt rushed to support its Arab ally.[25] Israel tried to explain that it had no hostile intentions toward Syria, but this effort was in vain. Other Arab states such as Jordan joined Egypt and Syria during the crisis, so Israel faced a possible Arab offensive on more than one front. Furthermore, Arab armies had much more armor than they had possessed in 1956, which increased their punch and their speed. This concerned Israel, mostly in areas where it lacked strategic depth, since the Arabs could have reached objectives inside those areas very fast.

The Syrian forces in the Golan were in an offensive deployment, but their activity during the war was relatively moderate, at least as far as starting an attack. Jordan did not do much, either. The Jordanian and the Syrian air forces conducted only a few sorties. The Jordanian and Syrian ground forces were no better. They kept busy firing on Israeli population, while Egyptian units fought to their bitter end in Sinai against the majority of the IDF.[26] The Jordanian and Syrian involvement was therefore negligible. Their air forces, as small as they were in comparison to their Egyptian counterpart, did not exploit the crucial moments in which almost all the Israeli fighter-bombers were sent to storm Egypt's airfields. The promises and treaties between the Arabs about mutual assistance proved meaningless. It was not a fine presentation of Arab solidarity in battle.

In 1973, Syria and Egypt recognized their deficiencies. After gaining lessons from past failures, the Arabs looked for reasons, not excuses, for the victories of the IDF in former wars. During this analysis they came to understand that even the most powerful Arab state was no match for the IDF if it stood alone. Joining forces and forming a tight coalition was essential for the Arabs. It was not enough to simply assemble the Arabs militaries, though. A military coalition had to act as one and not just strive for one general goal. The Arab alliance could be temporary, but it had to be stable and solid.

The 1973 showdown was quite an impressive demonstration of Arab unity.[27] They had several important advantages that every Arab coalition needed in combat. The Arabs had more troops and weapon systems than the IDF. Furthermore, in the first hours of the war, because of the Arab deception and Israel's wrong calculations, the IDF was enormously out-numbered on each front. Most of the IDF reserves were not ready to fight, or else they were on their way to the battlefield. This was the time in which the militaries of Syria and Egypt were on full alert and could have exploited such a rare opportunity.

At the beginning of the 1973 war the Arab coalition was formed only between Egypt and Syria, but they had two of the most powerful Arab militaries. They attacked Israel on two fronts that were hundreds of kilo-meters apart. This meant that each Arab military was on its own, and also that the IDF needed at least a few days to send large units (in an echelon of a brigade and higher) from one front to another. Therefore, during the entire war only half of the IDF was able to cope with all the Syrian and Egyptian forces.

The Arab coalition had to make compromises, such as on when to begin. Egypt preferred to start the attack at night because of its need to build bridges to cross the Suez Canal at a time when the IAF was less effective. Conversely, Syria wanted to begin the offensive in the morning, when the sun would blind the eyes of the Israeli troops in the Golan. Even-tually they agreed on attacking at 1400 hours (2:00 p.m.).[28]

Other Arab militaries contributed to the war effort—mostly Iraq, which sent about 500 tanks, and Jordan, which dispatched 170 tanks.[29] After Egypt and Syria, it was Iraq that contributed more manpower and weapon systems to the coalition than any other Arab state. Iraq's share, especially in the campaign near Damascus, was substantial. Iraq did not hesitate to dispatch an immense expeditionary force straight into the bat-tlefield. Its troops crossed hundreds of kilometers on their way to assist Syria. Their performance could have been better, but they did arrive at a very critical moment for Syria.[30] Iraq could have even helped more, but Syria and Egypt planned a surprise attack at the opening stage of the war, which emphasized secrecy. That, however, prevented serious coordina-tion—if any—with other Arab states prior to the showdown.

Three years before the 1973 war, Syria had invaded Jordan when the latter was torn in a civil war.[31] In the 1973 showdown Jordan was not eager, to put it mildly, to either help Syria or aggravate Israel. Jordan did eventually join Syria, out of fear of being accused of treason by failing the highest test of Arab solidarity: a war against Israel. However, the Hashemite kingdom managed to evade this frustrating and risky duty until the IDF's breakthrough into Syria.

During the clashes in the 1973 showdown, Arabs—mostly from the Jordanian, Syrian and Iraqi militaries—opened fire on each other by mistake, and they did not exploit their full potential.[32] This was partly because of the lack of coordination and partly due to operational confusion.

In the 1973 war, Syria was sure Egypt would move quickly toward the passes of Sinai: the Gidi and the Mitla that are located several dozen kilometers east of the Suez Canal.[33] But Egypt was satisfied with putting its flag over the east bank of the Suez Canal[34]—that is, capturing a few kilometers in Sinai. In spite of its alliance with Syria, Egypt focused on its own specific interest. Egypt's strategy was not purely a military one.[35] Its intention was not to throw the IDF completely out of Sinai, because it did not seem possible. For Egypt the military campaign was a vital, yet limited, stage, part of a political process—gaining back the peninsula by negotiations. Of course, if the postwar talks had failed to return to Egypt all of Sinai, Egypt could have tried to use its springboard in the peninsula in an attempt to conquer more areas there.

In 1973 Syria, which was worried about a drastic Israeli response to an Arab attack,[36] could have tried, instead of launching a massive offensive, to seize a tiny piece of land in the Golan,[37] such as a village near the border. Syria would have then held its offensive, while its patron, the Soviet Union, intervened and stopped Israel from retaliating. This step could have started a political process that might have ended in a complete withdrawal of the IDF from the Golan. Yet, in contrast to Egypt, Syria's military aim was not so limited. Syria wished to obtain in one major strike all the land that was taken in 1967—the Golan. This task was feasible, since the Golan is much smaller than Sinai. Syria was also aware that seizing the heights of the Golan would have been helpful in repelling the expected IDF counterattack, since Israeli troops would have had to climb uphill under heavy Syrian fire. Egypt would have enjoyed a more or less similar advantage if its forces had conquered the passes of Sinai. Those passes, because of their mountainous terrain, would have given the Egyptian units a comfortable position to push back Israeli attacks. Instead, Egyptian forces were exposed on the open ground of the east bank of the Suez Canal.

The Syrian anti-aircraft batteries covered all the skies of the Golan.[38] Egypt's anti-aircraft batteries were able to intercept planes only over a very small part of Sinai, which exposed Egyptian ground units to bombardments in the rest of the peninsula. This was another reason for Egypt to restrain itself. Indeed, Egypt, probably more than Syria, was aware of the weaknesses of its military. Even after its troops crossed the Suez Canal with relatively low casualties, Egypt did not change its strategy, in spite of Israel's military blunders during the first days. Apparently, after three failures in encountering the IDF in high-intensity wars (1948–1949, 1956 and 1967), Egypt was extremely careful. Storming over the Suez Canal, in almost complete surprise, and capturing isolated posts while the IDF was unorganized was a completely different challenge from confronting the IDF in maneuver warfare after its reserves had arrived. No wonder Egypt's army stayed in its bridgeheads, where it had more chances to stop Israeli assaults.

The full-scale attack Egypt launched on 14 October was carried out due to pressure by Syria.[39] But the IDF was ready; its tank cannons had longer range, and they were handled well.[40] The outcome was a fiasco for Egypt, since more than 200 of its tanks were destroyed, while the IDF lost about 25 tanks.[41] Furthermore, the initiative passed to the IDF, which exploited the opportunity to cross the Suez Canal and encircle the 3rd Egyptian Army. Although at the end of the war Egypt kept its grip in Sinai, according to its original intention, its gamble on 14 October showed the risks of exceeding its limited plans.

All in all, the brutal skirmishes and the cost of the 1973 showdown, together with the Arab successes, were a traumatic experience for Israel, with a lingering effect that fed the fear of losing a high-intensity war.

In the 1980s there were several options regarding how to maximize the Arab military potential against Israel.[42] By contrast, in 2012 there was not much chance of an offensive by several Arab states against Israel.[43] Even if another Arab coalition had formed, it would probably not have been as powerful as the one in 1973. Yet, as always, a massive surprise attack on more than one front could have brought Israel down.

# The Alliance Between Israel and France and Britain in 1956

In time of war Israel intended to rely on its own forces, and not on foreign powers.[44] This decision gave Israel more freedom of action. Yet, in 1956, in contrast to its other high-intensity wars, Israel collaborated with foreign partners—the European powers of France and Britain.

In the early 1950s Israel was worried, since Britain left Egypt military infrastructure such as airfields near the Suez Canal, which provided the Egyptian air force with a springboard to attack Israel. This could have encouraged Egypt to try surprising Israel from the air in a high-intensity war.[45] At that time negotiations between Britain and Israel about including the latter in a pro-Western alliance in the Middle East had failed. Britain preferred to reach an agreement with Egypt.[46] Yet, in October 1953, the planning branch of the IDF estimated that Israel could reach an understanding with Britain that would pave the way for Israel to conquer Sinai.[47] Furthermore, after Nasser nationalized the Suez Canal on 26 July 1956, Britain decided to overthrow him.[48] Britain thus joined France and Israel in planning an attack on Egypt.

On 12 September 1955 Egypt closed the Tiran Straits in the Red Sea for Israeli ships,[49] a blockade that could have been broken by the United Nations.[50] However, this organization, which had proved helpless in coercing Egypt to permit Israeli sailing in the Suez Canal,[51] failed to make Egypt obey international regulations in another sea route, the Tiran Straits. Western powers did not care about this problem either. It only bothered Israel. On the other hand, when Egypt nationalized the Suez Canal, the response of Britain and France was much quicker and more severe.

Israel concluded from this situation that expecting help from the United Nations or foreign powers depended on the degree to which the Israeli problem was relevant to their interests.

In 1956 Ben-Gurion, Israel's prime minister and minister of defense, feared that the Egyptian air force would launch a massive bombardment against Tel Aviv.[52] The IAF had a plan to target Egyptian aircrafts on the ground.[53] Yet Israel's general staff had doubts as to whether the IAF could implement such an ambitious task,[54] which had never done before.

Ben-Gurion, who was also not keen to rely on his pilots, demanded and received a commitment from France and Britain that they would neutralize Egypt's air force. Israel therefore trusted two foreign powers on a critical issue, since Egypt's air force could have bashed not only Tel Aviv but also Israeli columns in Sinai, vulnerable to bombardments.

Two days after the 1956 war began, France and Britain destroyed Egypt's air force.[55] Yet dozens of Egyptian IL-28 medium bombers managed to stay untouched until the end of the war by hopping between different airfields until French jets finally annihilated many of them. The rest fled to Saudi Arabia,[56] but until then they could have bombed Israeli cities.

In 1956, for lack of a better option, Egypt consented to losing land east of the Suez Canal—that is, Sinai—since Egypt's most vital sectors were west of the Suez Canal, such as the Cairo area,[57] Nasser's center of power, and the place where most of his troops were stationed. It could be argued that the outcome of this war was to a large extent decided outside the peninsula, since from its very beginning Nasser's main concern was securing the heart of his country.

Israel's convincing victory was due to the speed with which the 1956 war was conducted, coupled with the intervention of France and Britain that helped defeat Nasser and provided Jordan and Syria with a strong reason to stay out of the confrontation. Syria, which had been under the rule of France until 1941, did not wish in 1956 to give France an opportunity to gain back its control, as Britain tried to do with Egypt.

At the beginning of the 1967 showdown, after the Israeli air strike on Egypt, the latter claimed British and American planes participated in the attack.[58] Egypt's attempt to justify its failure to protect its airfields was based on the 1956 war. In fact, in 1967 Western planes did bomb Egyptian airfields, but they were French, not American or British. Furthermore, their air crews were Israelis.

## Israel and the United States

In 1950 the United States, France and Britain agreed to restrain the arms race in the Middle East by monitoring weapon sales to both the Arabs and Israel.[59] In 1956 the United States refused to sell Israel weapon systems.[60] Israel had to wait until the mid-1960s before buying from the United States the Hawk anti-aircraft missile,[61] tanks such as M-48 Patton[62] and fighter-bombers like the F-4 Phantom. The Hawk was more of a defensive system, while tanks and fighter-bombers could have been used in both offense and defense, which gave Israel more operational options. Yet, by relying on the United States in acquiring weapon systems, Israel lost some freedom of action.

Israel could have initiated a high-intensity war with the weapon systems it already had, but there was no guarantee that Israel would receive from the United States replacements for the platforms that would have been destroyed in battle. The United States could have reduced, delayed or stopped shipment of weapons, ammunition, spare parts, and so on (or threatened to do so) before, during or after a high-intensity war between Israel and the Arabs.

Overall, since 1962, the United States has given Israel military assistance in the amount of about $100 billion.[63] According to the United States

European Command, which deals with Israel, "commitment to Israel's security and well-being has been a cornerstone of U.S. policy in the Middle East since Israel's founding in 1948."[64] In 2013 "military and intelligence cooperation between the United States and Israel is considered to be better than ever. Moreover, despite political differences, cooperation between the two countries' national security councils is very strong."[65]

On the first day of November 1956, in the middle of the 1956 war, there was a demand in the UN assembly for a cease-fire. The United States opposed the war, and so did the Soviet Union.[66] On 24–25 October 1973, at the end of the 1973 showdown, the two superpowers agreed to stop Israel from obtaining victory.[67] This showed how sometimes Israel had to diplomatically confront both superpowers during a high-intensity war. On other occasions, mostly during the 1973 showdown, a mutual understanding between the Soviet Union and the United States served Israeli interests. A case in point was when the superpowers avoided joining the fight, thus preventing an escalation into an all-out global war.

The collapse of the Soviet Union allowed Western powers to act with more aggressiveness toward their rivals without fearing a global showdown.[68] The United States exploited this advantage in wars such as the 1991 and 2003 offensives against Iraq and the 1999 conflict in Yugoslavia. Israel benefited from the absence of the Soviet Union mostly in the wars against Iraq, particularly the one in 1991. Iraq was not a protégé of the Soviet Union, but even in that case the latter could not have intervened to block, or at least disrupt, this demonstration of American might.

During the war against Iraq in 1991 the United States dispatched Patriot missiles to Israel in order to intercept Iraqi long-range surface-to-surface missiles. Tactically speaking, the Patriot system failed, but strategically it succeeded in contributing to the American effort to stop Israel from joining the war in Iraq. An Israeli intervention could have undermined the fragile anti-Iraqi coalition, mostly because of its Arab partners. Israel retrained itself, since the United States and its allies, including (ironically) Arab states, had inflicted a major blow to the huge Iraqi military.[69] The victory of the Western-Arab coalition in 1991 in a classic high-intensity war reduced the chances of Israel having to face an Arab coalition. Iraq under Saddam Hussein was a threat to other Arab states, but Iraq was also a fundamental component in an Arab coalition against Israel. Without Iraq, such an alliance lost its most important expeditionary force, a convincing reason for why the Arabs ought not to clash with Israel again in a high-intensity war. The 1991 triumph of the anti-Iraqi coalition meant that the 1973 showdown was the last collision in which there was an anti-

Israeli coalition in a high-intensity war. Considering that in the first 25 years of Israel's existence there were Arab coalitions no less than four times (in 1948–1949, 1956, 1967 and 1973), the 1991 confrontation contributed to creating a period of more than 40 years (since 1973) without an anti-Israeli alliance, which presented the most dangerous challenge for Israel.

In 2013, from an Arab perspective, the United States had lost too much credibility to be considered a fair broker.[70] Yet the Middle East was "in enormous turmoil, crying out for a greater American role."[71] According to Michael E. O'Hanlon and Jeremy Shapiro in 2013, "Obama's priorities are clear: maintaining great-power peace, preventing the proliferation of weapons of mass destruction and combating Al-Qaeda and related groups." Maintaining stability in Arab states such as Syria and Iraq was less important.[72] Israel supported all those goals, although Israel's priorities were not the same as those of the United States. Stability in Arab states (mostly those that share a border with Israel) has been vital to keeping the peace with both Jordan and Egypt and the status quo with Syria. Regarding guerrilla and terror groups, Israel has been less concerned about Al-Qaeda or ISIS and more worried about the Hezbollah and Hamas. This shows the differences between the perspectives of Israel, a regional power in the Middle East, and the United States, a global power. Yet both of them have enough common ground in those matters so to allow them to cooperate with each other in a fertile and effective way.

## Unconventional Weapons

Israel has long been thought to possess nuclear weapons.[73] From the late 1960s onward all American administrations have accepted that Israel has the bomb.[74] Egypt has had chemical weapons (CW) since the late 1950s[75] and dropped this weaponry on its enemy in the war in Yemen in the mid-1960s.[76] The IDF was aware then that Egypt might aim this lethal arsenal toward Israelis as well.[77] This did not happen in the 1967 war, but Egypt's disappointment with its armed forces, which suffered a humiliating defeat, might have brought its leadership to explore other methods. In the 1973 showdown Egypt could have conducted an unconventional campaign. Israeli troops, who might not have been prepared for such a blow, one that they had never faced before, would have been stunned. Even those who would have survived the chemical attack might have been too confused and scared to respond effectively. In spite of that possibility,

Egypt avoided this approach, probably fearing a fierce Israeli retaliation. Using CW against Israeli targets in 1973 therefore remained for Egypt only the very last resort. Even after the IDF had crossed the Suez Canal, Egypt did not attack the advancing Israeli troops with CW.

It is possible that in the last days of the 1973 showdown Soviet Scud missiles with nuclear warheads were deployed in Egypt.[78] Yet this could not have forced Israel to accept Egypt's terms, since Israel had this kind of arsenal in its possession as well, and Egypt did not look for mutual assured destruction. Furthermore, the Soviet Union knew that any attempt to threaten Israel with nuclear weapons would increase the tension with the United States, which might have sparked a global confrontation.

On October 1990, Israel decided to distribute gas masks to its population, fearing that Iraq would strike Israel with CW.[79] This decision followed Iraq's conquest of Kuwait in early August of that year, which led to a crisis not only between Iraq and an international coalition but also between Iraq and Israel. There was reason for concern. In the 1980s Iraq dropped CW both on its own people and on Iranian troops as part of their eight years' war. Yet in the 1991 war Iraq did not use CW against the coalition or Israel. Iraq (or, more specifically, Saddam Hussein) probably feared nuclear retribution.

## Should Israel Have Launched a Bombardment Against Iran or Accept an Iranian Bomb?

"There are fundamental differences among mainstream Israeli thinkers about how to deal with Iran's nuclear program."[80] Israel could not have stopped Arab states from gaining CW, but it managed to prevent them from producing nuclear weapons when the IAF destroyed Arab nuclear reactors in Iraq in 1981 and in Syria in 2007. Israel also waged a long diplomatic campaign about stopping the Iranian nuclear program, hoping to convince the world to act in this matter while warning that Israel could attack Iran.[81] The United States, which has had a conflicted relationship with Iran since 1979,[82] could have attacked Iranian nuclear sites (or else allowed and assisted Israel in this mission).

Israel is intensely concerned about Iran's progress in its nuclear program, be it even a minor one. There has been a fierce ongoing public debate in Israel, which is often summarized in the following dilemma: Bombardment or a bomb? Should Israel launch an air bombardment on Iran or adjust to a new reality in which Iran has the nuclear weapons?

## 1. Bombardments and Bombardment

An Israeli attack could have started a long confrontation with Iran, with frequent mutual attacks, perhaps once a week or once a month. It would have been a war of attrition, obviously a painful and disturbing situation for Israel, but something the country has experienced and overcome in the past.

It was estimated in 2013 that a bombardment would delay the Iranian nuclear program for a few years at the most.[83] During an ongoing war with Iran, Israel could continue sending off sorties, particularly against the Iranian nuclear infrastructure. Even with no confrontation as a result of an Israeli bombardment, Israel could once again launch air strikes when Iran had rehabilitated its nuclear infrastructure. Either way, it could be an extended process in which Israel might need the involvement of the international community (or at least that of the United States).

Gulf Arab states—particularly Saudi Arabia—would prefer that Israel or the United States annihilate the Iranian nuclear infrastructure as soon as possible. It's likely, however, that Gulf Arab states would not openly support such a raid and would officially deny any collaboration with the attackers. Iran would probably accuse Gulf Arab states of perpetrating the raid in any case, in order to justify taking steps against them, including military ones. Hence, attacking the Iranian nuclear infrastructure might be too dangerous for the Gulf Arab states. Instead, they might seek an American nuclear umbrella to match a nuclear Iran.

Such raids by Israel would be a risky and complicated offensive; yet in the past the IDF has proved its ability to conduct daring operations. Implementing various military measures could have increased the chances for success. Those include all kinds of aircraft, long-range missiles and Special Forces, cyber warfare, and so forth.

## 2. Bombardments (by the United States or with American Support)

The Obama administration has declared its commitment to stop Iran from gaining the ability to launch nuclear weapons. Some in Israel hoped the United States would bomb Iran, either jointly with Israel or (preferably) without. However, it is doubtful that the United States would conduct a vast military operation against Iran, for several reasons.

The main challenge of the Obama administration has been the economy. A strike on Iran could have turned into an expensive war, jeopardizing the slow U.S. recovery from the Great Recession. The United States also wishes to avoid another war in Asia; with the war in Afghanistan winding down, few Americans wish to proceed to its neighbor, Iran. In addition, the United States managed for forty years to avoid a nuclear showdown with the Soviet Union, which was a strategic (if not existential) threat to the United States. The policy of containment, which was satisfactory enough against the Soviet Union, could be repeated against Iran.[84] From the U.S. perspective, the worst-case scenario would be Israel absorbing an Iranian nuclear attack, not least because of nearly 200,000 American citizens living in Israel.[85] However, Israel might survive, and, after all, the latter is not part of the United States.

Iran with a nuclear weapon would pose a serious problem for the United States but still far less than the Soviet Union did during the Cold War. Iran would undermine the position of the United States merely in the Middle East, while the Soviet Union tried to do that around the world. The Middle East itself might be less important to the United States in the upcoming decades, given its pivot to East Asia, rising oil and natural gas production in the United States and, in general, focusing on American internal issues. A nuclear Iran would therefore be a major concern for the United States, but not necessarily such that preventing it would be worth the risk of a war. The agreement on Iran's nuclear program makes this conclusion quite clear.

After assuming America would not attack Iran, and considering the possible ramifications of an Israeli attack, Israel at least wanted approval to carry out that mission itself.

Israel and the United States had to take into consideration all the factors as part of their preparations for the possibility of an Israeli attack on Iran. The United States wished Israel would not bomb Iran, since the latter could have retaliated against U.S. forces in the Gulf and other American targets and/or allies.

During an Israeli raid some of its air crews, who might have to bail out of their damaged aircraft near Iran, could have landed in the Gulf close to an American force. The latter could have saved them, but Iran would have been eager to capture those air crews. The result might have been a skirmish between Iranians and Americans that could have entangled the United States in a crisis—even a war with Iran. Rescuing Israeli air crews may seem like a minor issue, but it could have been one of the important ramifications of an Israeli raid that would have tested the relations between Israel and the United States.

The United States could have given Israel its biggest bunker-buster bomb, the GBU-57B, and a B-52 to carry it, since Israel's F-15 and F-16 could not do so. Having this advantage, Israel could have postponed its raid on Iran, because such an arsenal would have allowed Israel to crack Iranian sites when its current bunker-buster bombs could not have done this job anymore, since some Iranian nuclear sites would have been too fortified. Furthermore, delivering the B-52 and the GBU-57B to Israel would have bought the United States a lot of time, even years, to resolve the Iranian crisis in other ways.[86]

## 3. Israeli Bombardment
## But Iran Gets the Bomb

An Israeli raid might not have caused much damage to Iran's nuclear sites, but it would have provided Iran with greater motivation to increase its efforts in this field at any cost, even if Israel attacked again and again. Over the years Iranian leaders have expressed their hostility toward Israel.[87] An Israeli strike on Iran would have proved to many Iranians that their government was right—Israel is indeed a sworn enemy that must be confronted. One of the regime's claims might have been that although this time Israel bashed Iran with conventional ammunition, in the future Israel could use nuclear weapons. Such a possible danger would justify the necessity of Iran acquiring this deadly arsenal, so it could deter Israel.

An Israeli bombardment of Iran might have also been exploited by Iran to try and get rid of, or at least reduce, both the sanctions against it and the inspections of its nuclear sites. So an Israeli offensive against Iran might have helped Iran to produce a bomb.

## 4. To Live (But Not Love) the Bomb

Israel's strategy could start with avoiding a raid on Iran, hoping the Iranian people would overthrow their government before or after the latter could obtain the bomb. Failing that, Israel's remaining option would be to hope that the Iranian leadership, with a bomb in their hands, would choose mutual deterrence and not mutual destruction.

If there were a nuclear showdown, Israel might survive, but only barely. However, even without actually dropping the bomb, Iranian threats could cause a decline in immigration to Israel, along with major reduction in foreign investments, capital outflow, and so forth. Israel might not collapse, but it would face a serious challenge.

Russia helped Iran gain nuclear weapons capability by selling Iran technology in this field and blocking steps against Iran in the United

Nations. However, in spite of its soft cold war with the United States, Russia would not want Iran to possess nuclear weapons that would increase Iran's strength dramatically and improve its position among former Soviet republics—that is, Islamic states in central Asia and the Caucasus. This would be an unwelcome development for Russia, which considers those Islamic states its traditional sphere of influence. In spite of the cooperation between Iran and Russia against the United States in central Asia, they are still strategic rivals. Therefore, Iran with a bomb would be not only Israel's problem but that of others as well, including those that assisted Iran in getting nuclear weapons, such as Russia. They would be concerned if Iran felt strong enough to deal with them together with (or even instead of) focusing on Israel.

## 5. A Bomb in Exchange for a Bomb?

The Obama administration has put forth the idea of making the Middle East a WMD (weapons of mass destruction) free zone.[88] In spite of the turmoil in that area, some saw opportunity.[89] In the framework of an international accord, Israel could potentially give up its nuclear arsenal in exchange for nuclear disarmament of the Middle East as a whole—including Iran. But Israel would not agree to such a deal, even if it were possible to overcome the difficulties of monitoring this arrangement. Israel, fearing the Arabs, produced nuclear weapons in the 1960s, according to non-Israeli sources, long before Iran decided to be an enemy of Israel. Unless there is a genuine and comprehensive peace between the Arabs and Israel,[90] the latter will not surrender its last line of defense. It is also doubtful that Iran would abandon its nuclear weapons plans, which it needs primarily to protect its own regime, regardless of whether Israel has the bomb.

# 2

# Israel's Combat Doctrine in a High-Intensity War

There is a striking asymmetry between Israel and the Arabs in terms of population size, territory and natural resources, combined with lack of strategic depth on some of Israel's fronts. This makes one wonder about the odds of Israel's survival, which hitherto has depended upon the ability of its military to win major conventional wars, such as those that occurred in the era of Israel's high-intensity wars (1948–1982).

## *Taking the Strategic Initiative: Preventive War and Preemptive Strike*

In the 1950s, in case of an Arab invasion, the IDF could have commenced a counterattack. However, it was soon recognized that the lack of strategic depth could enable Arab armies to reach Israeli cities before the IDF was able to hold them back. This meant the IDF had to take the initiative from the beginning of the battles—that is, to launch a preventive war or a preemptive strike.[1]

It is possible that David Ben-Gurion, as prime minister and defense minister, thought about a preventive war against Egypt in 1951[2] and again in 1955.[3] In February 1955 there was an opinion in the planning branch of the IDF that Israel should be ready to launch a preventive war.[4] On 27 September 1955 Egypt signed an enormous arms deal with the Soviet Union, which served as one of the main reasons for the 1956 war.[5] Yet Egypt's military ran into difficulties in assimilating many of its new tanks and jets,[6] while during 1956 Israel acquired dozens of tanks and jets.[7] Israel assumed the IDF was assimilating its weapon systems faster and more effectively than its Egyptian counterpart,[8] although Israel increased its efforts to obtain more weapon systems after Egypt had already had a head start. Furthermore, Israel received fewer platforms than Egypt, while the latter had also the advantage of tanks, jets, and more of other Arab militaries.

In the late 1970s and early 1980s Syria was in the process of a massive military buildup, with the support of the Soviet Union.[9] In 1982 IDF's intelligence branch did not anticipate that Syria was about to attack Israel.[10] Nevertheless, Israel sought to reduce Syria's military strength.[11] In the 1982 war in Lebanon Israel might have wished to avoid fighting Syrian forces[12]; yet plans were made for such a scenario,[13] which was ultimately realized, possibly due to Israel's interests. Throwing Syria out of Lebanon was part of securing the north of Israel, by means of eliminating a potential threat to the Galilee. Israel also sought to increase its deterrence, since the growing military power of Syria might have encouraged the latter to retake the Golan. The Israeli offensive was supposed to stop Syria before it was ready to attack.

Israel's 1956 and 1982 confrontations were preventive wars.[14] They were not forced on Israel by the Arabs, as were the wars in 1948–1949, 1967 and 1973.[15] Israel's aim in 1956 and 1982 was to improve its strategic position in the Middle East and reduce the capability and determination of Egypt in 1956 and of Syria in 1982 to clash with Israel, at least for the foreseeable future. It had to do also with the tough collisions in 1948–1949 and 1973, when Arab states such as Egypt and Syria stormed Israel when the latter was unprepared for combat. Israel tried to make sure the next rounds—the ones in 1956 and 1982—would occur on its own terms as much as possible.

In February 1960 Egypt suddenly reinforced its troops in Sinai. Israel did the same, and for a few weeks there was a stalemate. It did not end in bloodshed, but rather with a mutual and quiet understanding to avoid a violent collision at that moment. The two sides decreased their deployment near the border.[16] Still, only four years after its defeat in 1956, Egypt had demonstrated its willingness to cause a crisis that might spark a high-intensity war.

For many years after the 1982 war, Syria concentrated much of its army, sometimes up to six divisions, on the Golan front.[17] Since the 1980s, Syria, like Egypt in 1960, after being beaten in its last clash with the IDF, gambled and provoked Israel by positioning a huge force near the Israeli border. This proved a certain failure of Israel's deterrence. Furthermore, in contrast to Egypt's decision to send a substantial part of its army into

Sinai only at the beginning of 1960, Syrian forces' deployment on the border was permanent. Their mission was to defend Damascus, although they could have also launched a massive surprise attack aimed at seizing the Golan. Yet, again like Egypt, Syria has not initiated a direct offensive since 1973, due to the memory of its former confrontation with the IDF (among other reasons). At least in that sense, Israel's deterrence worked quite well.

In 1982 major parts of Lebanon were under the direct control or influence of Syria, which regarded Lebanon as part of "Great Syria."[18] Lebanon served as a shield for Syria's western flank.[19] Yet the Golan, where 65 kilometers separated the IDF and the Syrian capital, was a more vital front. Lebanon, in the 1982 war, was therefore a secondary front for Syria, as Sinai was in 1956 for Egypt, which was focused on defending the Suez Canal against Britain and France. These constraints, which caused Nasser in 1956 and Assad in 1982 to allocate a relatively small part of their respective militaries to confronting the IDF in Sinai and Lebanon, helped Israel in accomplishing its goals.

The 1956 war started with an Israeli provocation: landing the 890th Paratroop Battalion deep inside Sinai.[20] In 1982 Israel waited for an Arab provocation to begin the war in Lebanon.[21] This opportunity came when an assassination attempt was carried out by a Palestinian organization against the Israeli ambassador in London. A large-scale operation based on that event was a non-proportional response; yet Israel exploited the attack in London to ignite a war in Lebanon.

In the 1982 war Menachem Begin, the Israeli prime minister, who was not familiar with military affairs, was "highly impressed by military power."[22] He believed that even if the reports of Arial Sharon, the defense minister, were not accurate, military actions were vital to Israel's national security.[23] Begin backed most of Sharon's moves, and it is not clear if Begin was deceived about some aspects of the war.[24] In 1956, Ben-Gurion, both prime minister and defense minister, trusted the top brass (especially Moshe Dayan, the chief of staff) in military affairs. Dayan exploited this trust, and sometimes he did not even update Ben-Gurion about major steps in the campaign, such as the decision to allow the 7th Armor Brigade to penetrate deep into Sinai.[25] The freedom of action that Dayan enjoyed in 1956, as did Sharon in 1982, enabled them to execute their plans. Nevertheless, the Israeli decision-making process in both of those confrontations lacked clarity and was riddled with disagreements.

In May 1967 Egypt reinforced its troops in Sinai.[26] The IDF, which since April 1950 has assumed it might have to conduct a preemptive attack,[27] implemented that maneuver in 1967. This option was also considered before the 1973 war but was not realized. In those two showdowns Israel was surprised by a sudden and massive deployment of an Arab army near its borders, crises that each ended in a clash. In 1967 Israel attacked, while in 1973 the Arabs struck first. In 1967 Israel recognized that taking the initiative in combat was justified even more than in a preventive war, since the risk was imminent—a matter of weeks, days and sometimes even hours. Waiting was risky, in case the Arabs charged first or had enough time to be sufficiently strong to foil the Israeli offensive. On the other hand, there was uncertainty about the intentions of Arab states. Had they actually wished to invade? Were they capable of succeeding? Israel also faced accusations of being the aggressor, as in a preventive war.

## The Time Factor

The IDF has always been aware of the significance of the time factor, exhibited in several key aspects both before and during a high-intensity war.

The IDF is based on reserves, which allows Israel possess the maximum amount of soldiers and, as a result, to narrow down Arabs' superiority in numbers.[28] However, the IDF needed early advance notice to have sufficient time to gather its reserves.[29] The crisis of May–June 1967 lasted for three weeks before the war started, allowing the IDF not only to collect its men but also to sharpen their skills.[30] In March 1973 the IDF was alerted about a possibility of war. In the upcoming months the IDF improved the readiness of some of its reserve units.[31] But when the battles actually began on October 6, the IDF received a very short warning of just a few hours, while most of its reserves were conducting their normal civilian life.

Therefore, it seems that while in 1973 the IDF had several months to upgrade its reserves, it had merely a few hours to amass them. In 1967 the IDF had only three weeks to organize and train the reserves, but most of them were already in active service when the war started.

The IDF has always focused on achieving a fast decision on the battlefield.[32] One of its main reasons for this goal is that Israel has always faced the possibility of a war on several fronts.[33] Israeli troops could have been moved from one sector to another by using internal lines.[34] Yet the numerical superiority of Arab militaries, particularly when they formed a coalition, could have forced the IDF to disperse troops on all the fronts. Israel feared leaving even one weak front for too long, lest it be the weak link broken by an Arab offensive. The IDF concluded it had to win quickly on one front in order to send reinforcements to another front, thus securing its success there as well.

About 6,200 Israelis, two-thirds of them soldiers, were killed in the 1948–1949 war.[35] No wonder that the IDF operations branch emphasized in the early 1950s the need to minimize the loss of human life during a war.[36] This meant striving for a fast victory, especially considering Israel's inferiority in the size of its population, contrary to the Arabs, who could drag Israel into a long and bloody fight.[37] Even a kind of draw in a high-intensity war, albeit with heavy losses to both sides, would have been a negative outcome for Israel.

In spite of the obligation to save lives, as long as the toll was not too high, the aspiration was to push speedily forward to beat the enemy. Therefore, ending the war rapidly did not guarantee fewer casualties. Continuing the war while waiting for the right opportunity to strike could have brought down the human price. But that option was often rendered impractical for Israel by other constraints, such as the need to discharge the reserves as soon as possible in order to prevent undermining the economy.

During the 1956 war, on October 30, the Egyptian top command finally understood what was happening in Sinai, but its efforts to stop the IDF failed because of the rapid pace of Israeli units.[38] One of them, the 27th Armor Brigade, rushed so fast into the north of the peninsula that some of its artillery could not catch up with its tank columns.[39]

In the 1967 showdown the IDF earned its triumph very quickly,[40] but in 1973, perhaps due to lack of time, the IDF did not gain a decisive victory.[41] Actually, in 1973 the Arabs fought better than in 1967, both strategically and tactically. When the tide started to turn against Egypt, and it was about to lose its 3rd Army, the superpowers intervened politically and stopped the IDF. However, on the Syrian front, following fresh and significant reinforcements from Iraq and Jordan, the end of the war came just in time for the exhausted Israeli forces there. A cease-fire, therefore, helped Israel on the Syrian front and blocked it on the Egyptian front.

## *Annihilating Arab Units and Conquering Arab Lands*

Two traditional operational goals of the IDF were destroying the adversary's military and conquering territories.[42] In the 1956 war three

Egyptian divisions disintegrated.[43] In the 1967 showdown Egypt's military lost up to 80 percent of its power.[44] Nevertheless, until October 1967 Egypt received from its patron, the Soviet Union, at least 80 percent of the weapons that were lost.[45]

Israel's interests in seizing Arab areas were many: increasing its depth, various economic and ideological reasons, and using the occupied lands as bargaining chips for the postwar negotiations. However, there was a price to pay for that achievement: enormous political and military difficulties for Israel in capturing huge populated areas, particularly cities like Cairo, and huge problems maintaining control there for a long period of time.

In the 1967 showdown, the IDF, while pursuing the crumbling Egyptian military, strove to block the vital passes of the Mitla and the Gidi located deep inside Sinai. The IAF bombed those sectors, but it was not enough. The Israeli ground units were required to halt the Egyptian retreat in those areas. Therefore, the task of capturing the passes combined the conquering of most of the peninsula and the destruction of the Egyptian military.

Before the 1973 war, the IDF had plans to capture vast areas in Syria and Egypt, such as the operation called "Desert Cat": seizing areas in the west bank of the Suez Canal.[46] In 1973 Israel had enough land—that is, sufficient strategic depth on the Egyptian front—since the nearest center of Israeli population was more than 200 kilometers away from the front line. Therefore, it was more important for the IDF to destroy Egyptian units than to conquer Egyptian territory.

In the 1973 showdown, in the Golan, Israel managed to repossess all the land that was taken from it during the first days of the war, and it further conquered 500 square kilometers inside Syria. On the Egyptian front the IDF seized large areas, but Egypt's military kept the territory that its troops had seized in Sinai. However, in spite of all the land and manpower Egypt and Syria lost in the 1973 showdown, the IDF was not able to coerce them into agreeing to recognize Israel's right to exist (similar to what happened in the 1967 war).

## The Size and Nature of the Battlefield

In the 1948–1949 war the IDF maneuvered with formations that had up to four brigades.[47] The IDF created the echelon of a division during the 1950s.[48] In the 1982 war the IDF had a corps that included a few divisions. But while the size of the units grew over the years, the battlefield did not; sometimes, it even became smaller. In the era of high-intensity wars the theater of operations spread, at least in theory, all over Israel, Egypt, Jordan, Syria, the north of the Red Sea and the Eastern Mediterranean, but in fact the main battlefield was often confined to much more limited sectors.

In the 1973 showdown, on the Egyptian front, thousands of soldiers and vehicles from both sides were squeezed into several dozen square kilometers near the Israeli bridgehead in the Suez Canal. In this limited zone the fate of the Israeli attack was decided, mostly in skirmishes in key crossroads and roads. The Egyptian military concentrated its fire on the bridgehead, inflicting heavy casualties among Israeli troops. After several days of fierce combat, IDF units stormed from the bridgehead into the west bank of the Suez Canal.[49] The main battlefield, which started in a relatively tiny part of Sinai, thus got even smaller when the IDF actually crossed the Suez Canal.

In 1973 one of the most vital posts Israel had in the Golan was on Mount Hermon. That base served as an extraordinary observation point, where one could scan a large part of the Golan. The Syrian

The Israeli bridgehead at the Suez Canal during the 1973 war. (Photographer: Ron Ilan; Source: Israel's Government Press Office)

military, very much aware of the importance of this location, made an effort to seize the site during the first hours of the war. The IDF tried almost immediately to retake that camp but failed. Only on the second attempt, at the end of the showdown, did the IDF succeed, suffering heavy casualties in the process.[50] The battles on Mount Hermon are an example of the significance of certain isolated spots in Israel's high-intensity wars.

In the years 1949–1967 the IDF trained its men in the south of Israel, in the open area of the Negev, preparing them to invade a similar landscape—the desert of Sinai.

In those same years, other exercises in the hills and mountains of the Galilee, in the north of Israel, helped Israeli troops cope with climbing up the narrow roads and paths leading to the Golan. The rugged terrain of northern Israel assisted the IDF again when training to maneuver in Lebanon in 1982.

Over the years, since the IDF fought again and again in the same terrain, its troops became familiar with it, a fact that contributed to their confidence and elevated their morale.

## Operational and Tactical Aspects

Israel's military doctrine was based on the offensive.[51] The IDF intended to gather its forces in a specific sector, in order to break through or infiltrate the Arab lines and penetrate deep into their land,[52] as it did in the late 1950s. In the 1956 war the 27th Armor Brigade stormed across the north of Sinai, about 200 kilometers[53] while the 9th Infantry Brigade traveled more than 200 kilometers to the south of Sinai and captured the vital port of Sharam al Sheikh.[54] In 1967 the 31st Division reached the heart of Sinai and stopped Egyptian reinforcements.[55] In the 1973 showdown, the rush into the Egyptian rear ended in the encirclement of the 3rd Egyptian Military. Deep penetrations had therefore all kinds of versions and goals.

Before the 1956 war Egypt fortified sectors near the Israeli border.[56] After the war, Egypt rehabilitated and upgraded its fortifications in the peninsula.[57] Nevertheless, the IDF paved its way into the peninsula once again in the 1967 showdown. The 84th Division charged on a main route, near the shore of the Mediterranean. At the Girady, on the road to the city of Al Arish, the elite 7th Armor Brigade faced a determined Egyptian resistance, although eventually this vital spot was seized by Israel. The

38th Division conquered another key position, in Um Katef, in a complicated maneuver that demonstrated cooperation between armor, infantry, airborne troops, artillery and engineers. Physical fitness was also necessary when Israeli reserve troops crossed 14 kilometers in the dunes of Um Katef on their way to mop up Egyptian trenches.[58]

The IDF was known to exploit opportunities during battle[59] while becoming acquainted with the Arab soldier and his weaknesses, such as lack of flexibility,[60] a trait that helped predict his behavior. For example, if an Israeli force pounced on an Arab post but failed to take over, the Arab garrison would many times recoil from attacking back. The IDF would then regroup and confront the objective again.

In the crisis before the 1967 showdown, the IDF tried to deceive the Egyptian military regarding the location of the Israeli attack, but it is not clear how successful this deception was.[61] Sometimes there was too much confusion on both sides, "the fog of war," as in the night between 15 and 16 October 1973, when Egyptian and Israeli forces collided on the east bank of the Suez Canal.[62]

Israel's Sherman tank operators, having to face superior Arab tanks, as the Egyptian T-34 in 1956,[63] tried to overcome their foe by implementing better tactics. It was one example of how, in the era of high-intensity wars, the IDF often had to be creative and develop various kinds of winning methods. This approach was intended to compensate for a lack in both quantity and quality of weapon systems. Israel, because of budget and/or political constraints, often could not afford to purchase advanced platforms.

## Protecting the Borders

Starting in the early 1950s, Israel mostly relied on civilian sites (towns, villages, kibbutz, etc.) as a defensive line in a high-intensity war.[64] This concept of territorial defense was based on lessons from the 1948–1949 war, when tough battles raged around the country. Sometimes the Israelis, while paying an enormous price, managed to halt or at least to slow down the Arab offensive. However, following the massive buildup of the Egyptian military since late 1955, depending on civilian posts might not have been enough in delaying—let alone stopping—Egyptian units on their way to the vital area of Tel Aviv.

Before the 1956 war Israel created fortifications in the south of the country.[65] Prior to the 1967 war, Israel did so again.[66] This reduced the exposure of its troops to Egyptian armor in the open terrain of the Negev. Yet Israel's posts were far from an unbreakable defense line. Israel could not have adopted a defensive strategy by building strong fortifications and obstacles along its borders—certainly not all of them, considering their length of several hundred kilometers. The idea of investing a large share of the military budget in such a project was dismissed because of its enormous cost.

In a high-intensity war Israel might have been able to afford to lose a few places near the border.

The main problem for Israel was estimating whether that was the first stage of a deep Arab penetration or just a limited move. This dilemma was especially prominent on Israel's central front, where an Arab offensive could have jeopardized the existence of the entire state.

Prior to the 1973 showdown, as part of defending the Sinai, the IDF planned to ignite the Suez Canal, which might have disrupted an Egyptian crossing. But the arson devices did not work.[67] Either way, the Israeli approach relied on attacking. Defense was considered a temporary step until the IDF began its offensive.[68] Therefore, at the start of the 1973 war Israel had an armor division in Sinai, ready to commence a counterattack. Its tank crews had frequently trained to reach the Suez Canal in the shortest time possible, in order to slow down and actually stop the establishment of an Egyptian bridgehead. It was forward defense that proved to have catastrophic consequences in the war, when Israeli troops could not prevent Egypt from seizing the east bank of the Suez Canal.

In 1973 Syria planned to recapture all the Golan. The IDF could have adopted defense in depth and, by this means, lured the Syrian military into a trap, thus buying time for reinforcements to arrive. Yet the Israeli troops were forced to apply forward defense.[69] They had near the border several posts inside knolls[70] that were used mostly for surveillance and observation. They were not supposed to be a formidable line of defense.[71] As in Sinai, the defense plan was based on the armor. During the war the Israeli tank crews had to stand firm and hold their ground under heavy shelling while facing superior numbers. In the northern Golan the 7th Armor Brigade held its ground, paying a heavy price for doing so, under continuing Syrian pressure. The results of the battle sometimes depended on a few Israeli tanks deployed in a key position, or else appearing from the flank to surprise the foe at the last moment. All in all, the bravery and resourcefulness of Israeli soldiers won the day for the IDF in the northern sector of the Golan. However, in the southern sector, in spite of fierce Israeli resistance, the Syrians broke through the line. The basic conditions there, such as the terrain, tended too much in favor of the Syrians.[72] Israeli hopes that forward defense would be sufficient were shattered on the Syrian front as well as on the Egyptian front.

One of Israel's lessons from the 1973 showdown was the need to improve and reinforce the territorial defense,[73] such as in the Golan. Settlements there, considering their proximity to the Syrian border, had to be integrated into Israel's defensive line. The IDF also upgraded its posts in the Golan during the subsequent decades.[74] The main burden continued to fall on the armor; yet the increase in the effectiveness of IDF firepower since the 1980s, mostly in the air, allowed it to inflict substantial damage. This advantage, together with the desire to reduce Israeli casualties, was a fundamental factor in dealing with another Arab offensive in a high-intensity war, particularly in the Golan.

## *Air Superiority*

The IAF strove for air superiority since the early 1950s in order to protect the skies over Israel.[75] Before the 1956 war, Israel bought the French fighter-bomber jet, the Mystere 4A. However, there were not enough qualified pilots for all the jets, and even those who did fly them lacked experience in handling the plane in air-to-air combat.[76] Fortunately for Israel, the Egyptian air force sent only two sorties into Israel's air space during the war, not causing any loss of life or great damage.[77] This was a minor Egyptian response to the Israeli invasion into the Sinai (then part of Egypt).

Before the 1967 war the IAF planned to neutralize Arab air forces.[78] On 5 June 1967 the IAF struck and basically destroyed the Egyptian and Jordanian air forces, while also causing heavy damage to the Syrian air force.[79] The IAF's amazing and successful knockout in 1967 left many Arab land units at its mercy, while the Israeli communication and logistic routes were secured from air attacks.

During the war of attrition in the late 1960s, Egypt became aware of the potential of the anti-aircraft missiles[80] and decided to rely on them more than on its planes.[81] Syria did the same. Although their jets participated in the 1973 war, it was mostly in the opening stage. During most of that collision the role of Arab planes was quite minor. If Egypt and Syria had invested more in their jets, their militaries would have had assistance in many situations in which the anti-aircraft missiles were useless. Unlike jets, the missiles were not able to intercept planes and also attack land objectives. Furthermore, if the anti-aircraft missiles had failed or been wiped out, the sky above the range of the anti-aircraft guns would have been wide open to the IAF. Egypt and Syria had no real backup to the anti-aircraft missiles in securing high altitude. The anti-aircraft missiles were not very mobile, either. Their batteries were similar to a phalanx—lethal but slow to maneuver. The anti-aircraft missiles had a killing range of a few dozen kilometers, while a plane could reach targets hundreds of kilometers away.

The constraints of the anti-aircraft batteries forced Egypt and Syria to deploy them near the front line, where they covered the land campaign. Since Egypt and Syria were determined to shield their troops in the main battlefield from air attacks, they had to deploy most of their anti-aircraft batteries in the Suez Canal and in the Golan, at the expense of protecting the Arab rear, including the capital cities, from Israeli air bombardments. This concept was particularly effective on the Egyptian front, since during the 1973 showdown in Sinai the IDF continued to rely on forward defense in spite of its disastrous consequences early in the war. As long as Israeli land units were stuck on the front line (without using the depth of the peninsula), the IAF had to support them, which exposed its planes to anti-aircraft fire. Yet Arab anti-aircraft batteries were vulnerable, too, because of their proximity to Israeli ground forces. Jets in air bases near the battlefield faced similar danger, but they could strike back against hostile land units advancing toward their airfields. Anti-aircraft batteries could not do the same, and they could not retreat as fast as planes, which were able to scramble in a short time if their airfield was about to be stormed by their foe. Egypt took those risks, and it suffered the consequences when Israeli tanks crossed the Suez Canal and destroyed the anti-aircraft missile batteries.

However, the Arab anti-aircraft formations, in spite of all their drawbacks, were still quite intimidating, due to their quality and quantity. The IAF absorbed many losses, especially in the first days of the 1973 war, when it had to support the totally outnumbered Israeli ground units. This constraint prevented the IAF from concentrating its power against the anti-aircraft batteries. In the upcoming weeks, the IAF learned vital lessons while contributing to the war effort. The IAF kept Israel's cities and most of the military infrastructure safe from air attacks, but many in the IDF believed the IAF could have done a better job—for example, against the anti-aircraft missiles.[82]

In the 1982 war, the IAF beat the Syrian air force and annihilated Syrian air defense batteries in Lebanon.[83] It was the third time in a high-intensity war that the IAF had to deal with anti-aircraft missiles, and the fifth time it encountered the anti-aircraft artillery in a high-intensity war (after 1948–1949, 1956, 1967 and 1973). (This summary does not include the war of attrition between Israel and Egypt in the late 1960s.) The IAF won this long struggle—at least in the last round—yet in 1982 it was quite a limited victory, since only a small percent of the Syrian anti-aircraft formation was in Lebanon. Actually, in 1982 the IAF had a bigger triumph over the Syrian air force when it lost about 80 jets in air-to-air combats.

## Strategic Bombardment

The IAF had difficulty providing close air support in all its high-intensity wars.[84] There were many complaints on this matter in the 1973 showdown.[85] In air interdiction the IAF gained more success, as seen in the 1956 and the 1982 wars.[86] With regard to strategic bombardment, in the 1956 war Israel wished to deter Egypt from dispatching its IL-28 bombers to drop their deadly cargo over Israel's civilian and military infrastructure. But the IAF had only a handful of B-17s, too few and too obsolete to run any effective campaign in the Egyptian rear. Egypt did not exploit this opportunity to bash Israel. In 1967 Israel did not possess any heavy bombers, but the destruction of Arab air forces paved the way for Israel to conduct strategic bombardment. However, Israel avoided that during the battles, since its planes were busy performing other tasks. Furthermore, Israel assumed its overwhelming victory would be enough to convince Arab states to end their conflict with Israel, and there would be no need to put more pressure on them by bombing their rear.

At the beginning of the 1973 showdown, Egypt made a few attempts to bomb Israel's population centers with long-range air-to-surface missiles. Those projectiles missed or were comparative slow, which helped Israel's jets shoot them down.[87] The IAF, for its part, could have launched strategic bombardment against Egypt's cities with relative ease.

This freedom to perform in the air was due to two basic factors: the IDF's superiority over Egyptian fighters trying to intercept Israeli planes, and the fact that most of Egypt's anti-aircraft missile batteries were deployed near the Suez Canal, quite far from its major cities. It was also more challenging for the IAF to identify camouflaged targets on the front line than a building or a factory in an urban area.

In 1973 Israel conducted an air offensive against Syria's rear, as a response to the latter firing Frog rockets on a kibbutz.[88] Israel also sought to break Syria's will to fight by commencing an attack both on the ground (from the Golan) and in the air (a strategic bombardment). Success in this arena would have allowed Israel to focus on Egypt, but Syria held on even after its general staff building, airfields, base depots, refineries and power stations, were struck by air attacks. Israel's campaign against Syria achieved limited results on the ground as well as in the air. Israel could have continued, and even intensified, its efforts in the air, but its intention was not to cause an escalation into an all-out war, in which the Soviet Union might have intervened. Israel was less concerned about a Syrian retaliation. Syria's planes had little chance of penetrating Israel due to the latter's advantage in air-to-air combat, and Syria's Frog rockets had a relatively short range.

Nine years later, in the 1982 war in Lebanon, Syria had surface-to-surface missiles, which could have reached the Israeli rear. It was one of the reasons why the confrontation in 1982 did not spill over into Syria—that is, to another strategic bombardment there.

## Anti-Tank Weapons

The new tank the Arabs assimilated before the 1973 war, the T-62, proved to have the same vulnerability as their veteran T-54 tank.[89] Yet Egypt's army had more tanks than the IDF.[90] The Arabs had also anti-tank missiles of which the IDF was aware, so running into them was not a total surprise.[91] Still, in the 1973 showdown the biggest challenge of the Israeli armor was not Arab tanks, as in 1956 and 1967, but rather the anti-tank missile and the anti-tank rocket. The Israeli armor was not ready to deal with a massive use of anti-tank weapons by the Egyptian infantry. The Israeli armor trusted its tanks to do whatever was required without much (or even any) support from Israeli artillery and infantry. The IDF stuck to a concept that failed, since its tanks were not versatile and protected enough.[92] The Arabs combined the new anti-tank missile with more traditional anti-tank measures: artillery, mines, and so forth. Together they bashed the Israeli armor and prevented it from gaining another success like that of 1967.

Regardless, the Israeli armor had some advantages over the anti-tank missile. The range of the latter was just a few kilometers, so it could not annihilate targets that were not in its line of sight. The tank had a longer range, and it could hit objectives with indirect fire. The tank had much better mobility and speed than the anti-tank missile when it was carried by the infantry, a substantial factor in an open area such as the Sinai. The tank, because of its armor, was also protected all the time, while an infantryman with the anti-tank missile had to look for cover, being completely exposed. The main weapon of the tank, its cannon, was effective against both tanks and infantry, while the anti-tank missile was mostly meant to attack tanks. The tank also had heavy machine guns against infantry. So Egypt's dependence on anti-tank missiles and not on armor seriously limited the scale of its operations not only in the tactical dimension but also in the operational one.

## Naval Warfare

The Israeli navy had to watch Israel's coasts and sea routes in both the Mediterranean Sea and the Red Sea.[93] In the 1950s and 1960s Israel imported oil by sea from states like the Soviet Union and Iran.[94] Israel depended on its sea trade because of the hostility of Arab countries surrounding it, preventing any passage of materials, products, and so on by land to or from Israel. If Arab navies had enacted a blockade for a long period of time, they could have deprived Israel of essential resources. Such a crisis would have crippled the Israeli economy, possibly even grinding it to a stop. Israel would have been brought to its knees.

In the 1973 showdown Israel's naval strategy was aggressive, but its naval doctrine and new missile boats, along with various other measures, were not tested in combat. The sea clashes occurred mostly close to the shore. Sometimes coast artillery joined the skirmishes between vessels of both sides. After a few days of fighting, the Arab fleets absorbed serious losses and withdrew to their ports, even hiding behind civilian ships. The swift Israeli victory at sea foiled the Arab ambition of cutting Israel's sea lines in the Mediterranean Sea.

However, keeping Israel's sea routes open to East Asia and Africa by sailing through the Red Sea was a different matter. Egypt closed the southern gates to the Red Sea, the Bab El Mandeb Straits.[95] The IDF could not destroy the Egyptian ships in Bab El Mandeb, and it certainly could not stay there to keep those straits open, due to their distant location, about 2,500 kilometers from Israel. Another option for Israel was to bomb Egypt's rear, or at least threaten to do so, as long as Egypt continued with its blockade in the Red Sea. Yet Israel avoided making such a move. Israel did not respond by disrupting the transport of supply by sea from the Soviet Union to Egypt, either, for that might have led to a furious Soviet retribution, and thus to a globally explosive situation, since the United States might have gotten involved too.

In the 1956 war the Israeli navy assisted the 9th Infantry Brigade in its task: seizing the port of Sharam al Sheikh in the south of Sinai. One form of aid was providing the brigade with AMX-13 tanks, thus saving them a voyage of hundreds of kilometers, partly on the tough sandy terrain of the Sinai desert.[96] A much bigger landing was planned before the 1973 showdown, on the west bank of the Suez Bay. The concentration of Egypt's military along the Suez Canal left its coast in the Suez Bay exposed. A daring amphibious assault there could have brought the IDF to Egypt's rear[97] without paying the price of breaking through the Egyptian defensive lines. Such a penetration could have created an opportunity to encircle a large part of Egypt's military, thus changing the course of the war. Yet there were no Israeli missile boats in the Red Sea to protect the slow and vulnerable Israeli landing craft from Egypt's missile boats. The IDF also hesitated because landing from the sea was a complicated maneuver involving the land, sea and air, for which its brass had no operational experience. Indeed, until the end of the war there was no amphibious invasion in the Red Sea, which kept the battles confined to the Suez Canal.

Following Israel's victory in the Mediterranean Sea in 1973, the conditions were set for a major amphibious operation there, but Syria's shores were far from its centers of government and industry. Vertical flanking there might not have produced the right effect. An Israeli bridgehead in Lebanon would have been much closer to Damascus and other major cities in Syria, while avoiding Syrian defense disposition in the Golan. Either way, Israel's landing craft were in the Red Sea, and Israel could not have transferred any vessels between its two naval arenas unless they circumnavigated all of Africa, which would have taken too long.

# 3

# The Strategic Linkage Between Israel's High-Intensity, Hybrid and Low-Intensity Wars

In the years 1948–1982 Israel had a series of low-intensity wars that were actually border wars with Arab states: Egypt, Jordan, and Syria, as well as a non-state actor—the Palestinians. Those collisions started with infiltrations of Palestinians into Israel in the 1950s, bringing Arab states into these problems. The low-intensity wars continued in the 1960s between Israel and the PLO, Jordan and Syria. In the 1970s and up to 1982 the PLO took center stage when it clashed with Israel. Since the 1980s Israel has been focused on low-intensity and hybrid wars with the Hezbollah and the Palestinians, including the Palestinian Authority (PA).

There were several types of strategic linkages between the high- and low-intensity wars, particularly the way each of them caused the other to happen. There were also scenarios in which low-intensity wars might have been realized following a high-intensity war and vice versa, such as a mutiny inside an Arab state after its defeat in a high-intensity war with Israel. Another aspect is comparing invasions in low- and high-intensity wars.

## Israel Versus the Palestinians, 1949–1956

In the 1948–1949 war, about 600,000–760,000 Palestinians became refugees.[1] They left behind their property in what has since become the state of Israel and were settled in camps in Arab countries. The Palestinians wished to return to their homeland, and some of them wanted to do so right away, in spite of the dangers they faced. Naturally Israel, as a sovereign state, strove to protect its borders, infrastructure and population. The result of the 1948–1949 high-intensity war was therefore an ongoing low-intensity war between Israel and the Palestinians.

The main threat to Israel was a high-intensity war, in which Arab states launched a full-scale attack. The Palestinian infiltrations were also a kind of invasion, since they penetrated the same fronts from which Israel expected an Arab offensive (the West Bank and the Gaza Strip) while enjoying the same advantage (the short distance from the border to Israeli towns due to Israel's lack of strategic depth). However, the Palestinians did not possess a conventional force with armor, artillery and bombers. They instead came in small groups, even as individuals, walking on foot.

Palestinians infiltrated Israel because of economic reasons, among others.[2] Economic goals likewise played a part in high-intensity wars against Israel—for example, Jordan sought an access to the Mediterranean Sea. In comparison with Arab states, the aim of the Palestinians (to provide for themselves and their families) was much more modest. Yet, for Israel, the unlawful crossing of Palestinians into its territory constituted a threat. Although many Palestinians did not necessarily wish to encounter Israelis, especially members of the police or the military, some Palestinians did see confronting Israelis as a primary—if not the only—purpose, one worth risking their lives for.

Arab states could have grabbed Israeli land following a high-intensity war serving as a springboard for more attacks on the rest of the country. A permanent Palestinian presence on part of Israel's territory, as a result of penetrations during the low-intensity war, might have gradually undermined Israel's right to exist. In such a situation, the Palestinians would have been encouraged to expand their activity to other sectors while sending more people in.

Occupying areas inside Israel during a low-intensity war would have taken longer than conquering them in a high-intensity war, but this approach had its advantages. The world might have considered it a natural process whereby Palestinians were returning to homes they had lived in for centuries, particularly if the Israeli resistance to that process was weak and vague.

Egypt initiated the 1973 high-intensity war as a way of attracting international attention to its demands. The Palestinians, by conducting a low-intensity war in the 1950s, wished to do the same, letting other nations know that the Palestinians had not given up their national dream. From their perspective, they were only returning to their property, sometimes by using force that hurt Israelis. The Palestinians saw themselves as victims trying to repossess what was stolen from them by Israel—and, if need be, doing so through a minor invasion.

The Palestinians did not mold their actions into a unified strategy, but it was enough to disrupt the day-to-day life in Israel. The general idea was that if the Palestinians could not have their way, the Israelis would not be permitted to relax and enjoy their state. Gradually many Israelis felt they were not safe in their own state, although providing security for Jews was one of reasons why Israel was established. The Israelis were aware that their region was a tough neighborhood, but realizing that even in their own private homes they were at risk all the time was too much. It was bad enough accepting that their entire state, including their homes, could be jeopardized in a high-intensity war.

The early to mid-1950s came only a short time after a high-intensity collision, the 1948–1949 war. The fighting could resume, even at short notice, but the high-intensity war was not seen as an immediate threat as much as the low-intensity war—that is, the Palestinian incursions. For Israelis, it was a grim reality. Other Jews from around the world might have hesitated to immigrate to a dangerous country. Foreign investors might have stayed away as well. The Palestinian infiltrations could therefore destabilize Israel, which was already burdened with many serious economic and social problems.

In the years 1949–1967 the IDF could have faced an Arab rebellion, following the conquest of Arab territory during a high-intensity war. A mutiny might also have happened inside Israel after the 1948–1949 war, when Palestinians came under Israel's direct control. But such an uprising did not occur in spite of all the violent confrontations between Israel and the Palestinians who infiltrated from neighboring Arab states. The Israeli Palestinians stayed out of this clash, and by doing so they avoided starting a low-intensity war of their own.

## The Role of the Arab States, 1949–1956

Ever since the 1950s the IDF had to be ready for different types of high-intensity war scenarios.[3] For Israel, the cost of a total defeat in a high-intensity war might have been the loss of its independence, if not worse. Under such circumstances, Israel would probably have been divided between the Arab states. Those Jews who had chosen to stay in the country, and were allowed to do so, would have been under foreign rule—an Arab one. In the past many Jews have been subordinated to Muslim regimes for more than a millennium, and they have known life as a minority in Arab countries. Yet, after experiencing life as a majority in their own state, losing that position in a high-intensity war could have caused some of the former citizens and soldiers of Israel to become insurgents. This would have been a low-intensity war between them and the Arab army and police in charge of keeping order in the conquered territories.

The failure of Egypt in the 1948–1949 war was one of the reasons for the resentment toward its monarchy that brought about the revolution of 1952.[4] It was quick and bloodless, since the Egyptian king did not possess loyal troops to crush the mutiny. Naturally, Arab regimes became aware that losing a high-intensity war with Israel might endanger their authority, or else provoke a low-intensity war (i.e., a rebellion) within their countries. They realized that in the next high-intensity war with Israel they must keep part of their troops to secure the government, even at the expense of the war effort against Israel. But a shortage of men on the front line could bring defeat, with all its ramifications for the stability of the Arab regime. This constraint could have encouraged Israel to initiate a high-intensity war, with one of its goals being to stir up the Arab populations and/or militaries against their governments. Such a consideration was based on the belief that peace with the Arab states could only be achieved if their regime were a democracy.

As it happened, the penetrations of Palestinians into Israel from neighboring Arab states led to talks between Arab governments and Israel, such as in the joint armistice committees. Arab states sometimes tried to reduce the amount of infiltrations, but it was not enough.[5] Generally speaking, Arab states were not bothered too much—if at all—about Palestinians wreaking havoc inside Israel. Yet Jordan often made an effort to stop Palestinian infiltrations, since it was worried about an escalation of border wars into a high-intensity war, which could have resulted in Israel invading the kingdom. This would have turned Jordan into a battlefield between itself, Israel, the Palestinians and perhaps other Arab militaries as well, mostly from Iraq and Syria. Jordan might not have survived such a confrontation. The 1948–1949 war had already demonstrated the sensitivity and susceptibility of Jordan to movements of other Arab militaries in its territory.

At the beginning of the border wars in the 1950s, Israel's government tended to deny that the IDF conducted raids on Palestinian villages in response to Palestinian infiltrations. Israeli leaders instead placed the responsibility for those actions on vigilantes.[6] Israel wished to avoid an escalation that might have ended in an unnecessary high-intensity war, and so the government described the assaults on Palestinian villages as local clashes, a kind of a private border war between the populations on both sides. This was a very vague explanation, if not a ridiculous one. It looked like Israel, similar to the Arab states around it, pretended that it did not control the series of skirmishes between Palestinian and Israeli civilians. Both the Israeli and the Arab leaders were well aware that this was a border war between the Palestinians and the state of Israel—that is, its government.

Until October 1953, Israel retaliated against Palestinian villages, but since that time the Israeli targets have changed to Arab military and police camps.[7] The aim has been to put pressure on the Arab regimes in Egypt and Jordan to increase their efforts in reducing the scale of infiltrations into Israel. The border war between Israel and the Palestinians now included border wars between states. The collisions between Israel and Egypt or Jordan were quite limited, but they raised the probability of a deterioration that might escalate into a high-intensity war.

After the 1952 revolution in Egypt, Israel hoped the new regime would negotiate with it,[8] but in spite of several contacts between the two states, no agreement was reached. The former ruler of Egypt had failed to destroy Israel in 1948. The new Egyptian regime continued the policy of its predecessor, refusing to recognize Israel as a legitimate state. However, because of other priorities and constraints, such as the current condition of its military, Egypt avoided another attempt to annihilate Israel in a high-intensity war, even if joined by other Arab militaries. It would therefore seem that Egypt should have done more to stop Palestinians infiltrations and a possible border war that could entangle Egypt in a high-intensity war.

Indeed, up to 1955 Egypt made some attempt to block penetrations from its territory into Israel,[9] but it also dispatched reconnaissance patrols from the Gaza Strip, disguised as Palestinian infiltrators.[10] Furthermore, during 1955 Egypt organized the "Fidayun"—the ones who are willing to sacrifice themselves—from among Palestinians in the Gaza Strip. They were trained and sent to carry out guerrilla and terror assaults inside Israel.[11] The creation of this semi-military unit increased the tension in the ongoing border wars (and thus the chances of a high-intensity war between Israel and Egypt).

Moshe Dayan, as IDF chief of staff, who carried a lot of political weight, strove for a major collision with Egypt erupting from the border war.[12] This is why Dayan continued the raids on Egyptian posts and even expanded them. The shift in Egypt's strategy, the creating of the "Fidayun," played into his hands. Yet a high-intensity war was a gamble, and Israel could have lost.

Israel and the Arabs looked for signs suggesting their opponent was preparing for a high-intensity war, such as starting a massive military buildup. Israel could have absorbed Palestinian infiltrations, but an attack from several well-armed Arab divisions was a much bigger threat. From the Arab perspective, they could have afforded a temporary seizing of their posts during Israeli raids, but a full-scale offensive of the IDF, after being re-equipped, was a completely different matter.

On 27 September 1955 Nasser declared that his country was about to receive more than 100 jet fighters, 50 jet bombers, and hundreds of armed vehicles such as tanks.[13] It was possible Egypt wanted that arms deal because of an Israeli raid in Gaza that occurred in late February 1955.[14] Others disagree.[15] However, even if one skirmish from the border wars was not the reason for Egypt's massive military buildup, the result was still a major improvement in Egypt's ability to launch a high-intensity war. Egypt was on its way to turning its military into an armor one, backed by a mighty air force. Egypt also could have enjoyed superior numbers of infantry units, thanks to its vast population.

Israel had to accept a new strategic reality in which its greatest foe, even without the support of other Arab militaries, was about to obtain an overwhelming military advantage. This situation made the border wars a danger of escalation to a high-intensity war, and it could have changed Israel's approach, but during 1956 Israel, having received hundreds of tanks and dozens of jets from France, apparently felt strong enough militarily to go on with its raids against Egypt.

Egypt blocked the Tiran Straits on 12 September 1955, which cut Israel's sea routes in the Red Sea. Israel's response was to consider a large operation called "Omer," which aimed at breaking Egypt's siege in the Red Sea.[16] This action would have probably caused a high-intensity war between Israel and Egypt, so the Israeli government abolished "Omer" in early 1956. In spite of the value of the Tiran Straits, Israel was not ready to go to war solely for this reason.[17] Indeed, "Omer" was a much different and bigger move in comparison to Israeli raids during the border wars, and that fact emphasized the risks and ramifications of this operation.

As to Israel and Syria, they had a limited, albeit extended, border war in the 1950s. Its peak was the battle on Tel Motila in May 1951 and the Israeli raid in December 1955 on several Syrian posts that were on the east bank of the sea of the Galilee. Those skirmishes increased the tension between the two states, but they did not bring about a high-intensity war.[18]

## The Role of Foreign Powers, 1949–1956

The border wars between Israel and the Arabs in the 1950s might have interfered with the intentions of Western powers to create an anti-Soviet alliance in the Middle East. Hence, a low-intensity war between Israel and the Arabs could have disrupted the preparations of Western powers for a high-intensity war with global dimensions.

In the early 1950s Egypt tried to get rid of the dominating British presence.[19] In 1950–1951 British forces in the Suez Canal faced a series of guerrilla attacks.[20] This low-intensity war would have increased the chances of a high-intensity war between Israel and Egypt if British troops had left Egypt and ceased to be a kind of a buffer between Israel and Egypt. Although the British deployment in Egypt did not prevent the Egyptian attack on Israel in 1948, the British departure from Egyptian bases in the mid-1950s might have raised the probability of a high-intensity war between Israel and Egypt.

The escalation in the border wars between Israel and Jordan in 1955 caused some concern for Britain, being in a way Jordan's patron. Britain even had contingency plans to bomb Israeli airfields.[21] This situation demonstrated how a low-intensity war between Israel and an Arab state might have evolved into a high-intensity war between Israel and a European power. However, the low-intensity war between Israel and Jordan continued, and in 1956 Britain not only avoided attacking Israel but also joined it in a secret agreement, which included France as well, against Egypt. Therefore, the tension between Israel and Britain during the low-intensity war between Israel and Jordan did not prevent Britain from collaborating with Israel in a high-intensity war against Egypt.

Israel started the 1956 war by landing a battalion deep inside Sinai. According to its pact with France and Britain, this should have given them an excuse to invade Egypt. If the two European powers had withdrawn from their obligation toward Israel, the latter could have pulled out its troops and presented the entire operation as just another raid, albeit a large-scale one.[22] In this case, what eventually turned out to be the first stage of a high-intensity war would have seemed like another border war battle.

In 1956 Iraqi troops might have entered Jordan to assist it against Israel during the border wars.[23] Israel, wishing to avoid a confrontation with Britain or Iraq, did not want to divert the regional attention to the Hashemite kingdom so that it jeopardized Israel's focus on Egypt.[24] Therefore, the border war on the Jordanian front was a constraint for Israel when it was planning a high-intensity war against Egypt.

In the 1950s Nasser aided rebels in Algeria against France. The latter believed that bringing down Nasser would help the French forces overcome the mutiny in Algeria.[25] By joining Israel and Britain against Egypt in 1956, France assumed that a high-intensity war against one Arab country, Egypt, would help end a low-intensity war in another Arab country in North Africa—namely, Algeria.

During the 1956 war, Israeli troops watching enemy movements on the Syrian and Jordanian fronts saw almost no sign of Arab solidarity. Perhaps Syria and Jordan, which, in contrast to Egypt, had not received hundreds of tanks and jets, were unsure about the capabilities of their militaries in a high-intensity war. A hasty gamble might have ended in their defeat, thus raising the chances of a low-intensity war at home—that is, a rebellion against the local regime.

## The Border Wars in 1957–1967

Although Israel was compelled to retreat from the Gaza Strip in early 1957, it is possible that the withdrawal saved Israel from enduring a low-intensity war. Insurgents could have confronted the IDF in the Gaza Strip as part of Egypt's attempt to gain back its lost territory. After all, Egypt had already used the Palestinians there against Israel during the border wars before 1956.

One of Israel's goals in the 1956 war was to destroy the infrastructure of terror and guerrilla warfare in the Gaza Strip.[26] The IDF accomplished this mission,[27] and the border between Israel and Egypt, including in the Gaza Strip, was relatively quiet in the upcoming years, at least until 1962.[28] In this sense, a high-intensity war stopped a border war. All the Israeli strikes on this front until 1956, during the border wars, were not as effective as one large offensive in a high-intensity war.

Before the 1956 war, Egypt, fearing an Israeli offensive, reduced the "Fidayun" assaults from the Gaza Strip while continuing to support such operations on the Jordanian border.[29] Nevertheless, Israel started a high-intensity war on the Egyptian front in order to prevent Egypt from resuming the "Fidayun" activity—that is, the border wars.

In February 1960, after an Israeli raid in Syria, the Egyptian army suddenly deployed hundreds of tanks in Sinai. The crisis, which was caused by a low-intensity war between Israel and Syria, could have created a high-intensity war between Israel and Egypt. The affair ended peacefully; yet clashes on the Israeli-Syrian border continued during the 1960s over control of demilitarized zones and water sources of the sea of the Galilee. Those border wars could not only have ignited a high-intensity war between Israel and Syria but also dragged Egypt into it.

During the 1960s the PLO hoped its activity against Israel, such as laying mines and explosive charges on the border, would cause a high-intensity war between Israel and the Arab states.[30] Israel retaliated against Arab states from which the PLO operated, in order to convince their regimes to halt this low-intensity war. Jordan, wishing to avoid an escalation into a high-intensity war, attempted to stop Palestinian assaults from its land, but sometimes its effort was insufficient.

On 13 November 1966, the IDF conducted a raid on a Jordanian village, Samoa,[31] after an incident in that sector in which a mine killed three Israeli soldiers. The Jordanian legion intervened, and the two militaries clashed with each other, although on a very limited scale and for few hours only. Palestinians, who wished such a skirmish would ignite a major collision, were disappointed. After this battle Jordan could have mustered its forces against the PLO, preventing it from provoking Israel. However, allowing the Palestinians to confront Israel lowered the chances that they would turn their frustration against the Hashemite rule, an unrest that could have become a rebellion. Already, following the Israeli incursion, there were Palestinians who started riots and threatened the stability of Jordan.

Those possible ramifications demonstrated how a low-intensity war between Israel and Jordan might have caused a high-intensity war between them or a low-intensity war inside the kingdom. For Israel, with all its problems due to the border war with the Palestinians, striving for a mutiny inside Jordan was not a better alternative. A collapse of the Hashemite kingdom would have paved the way for a new government, which might have been much more hostile toward Israel. This could have led into deterioration in the border wars and even ended in a high-intensity war. Thus, for Israel the border war was the lesser of two evils.

## The 1967 War

Egypt did not restart the border wars in the years 1957–1967, since they could have sparked a high-intensity war and its military needed time to prepare for such a confrontation. Egypt was also heavily involved in a civil war in Yemen. This low-intensity war confined many of its units, up to 70,000 troops,[32] and reduced the chances of Egypt seeking a high-intensity war with Israel.

It is possible that Nasser wished to destroy Israel in the 1967 showdown.[33] Indeed, the traditional ambition of Egypt was to annihilate, or at least weaken, Israel as much as possible, as part of the power struggle in the region. Egypt challenged Israel directly by both concentrating its army in Sinai and closing the Tiran Straits on 22 May 1967. However, it is not certain whether Arab leaders—particularly Nasser—wished to replace the border wars against Israel (those taking place on the Jordanian and Syrian fronts) with a high-intensity war. Perhaps the Arab radio and newspapers proclamations from that time should be separated from the realistic political opinions of Nasser and other Arab leaders. Maybe Nasser exploited the crisis to pull back his troops from Yemen, and the prospect of a high-intensity war with Israel was just an excuse to end a problematic low-intensity war in an Arab state.

In the days of May 1967 there was a fear in Israel of a clear and present danger[34] in spite of the successes in high-intensity wars in 1948–1949 and 1956 and in low-intensity wars since 1949. For the Israeli public, there was a growing probability of a costly and difficult confrontation like the one that occurred in 1948–1949, which reminded the Israelis once again that a complete defeat of the IDF might put an end to their state.

Israel's main strategic aims were to break the blockade on the Tiran Straits and to beat the Egyptian military and, along with it, the Arab coalition.[35] Although Israel had to be ready to fight on several fronts, Sinai was the most essential one, because a defeat of Egypt might have pushed other Arab states to abandon the coalition. Israel also planned to conquer Arab areas, thus preventing a wave of guerrilla and terror attacks from there.[36] Israel's goals therefore illustrated the need to win in a high-intensity war while preventing a low-intensity war.

Egypt, Jordan and Syria had all kinds of disputes with each other.[37] For Jordan, Israel has been an official enemy, but not the only one, and not even the most dangerous one. From the Jordanian perspective, Egypt also represented a threat, particularly in circumstances such as those that prevailed in 1967, which were used by Nasser to increase his influence in the Arab world. This friction between Arab states reduced their will to exhaust their capability to achieve one main goal—defeating Israel. The Arabs could have pinned down large Israeli units within different sectors.

At least at the beginning of the war, Syria, Egypt and Jordan did not have to commence a major offensive—they only needed to pretend to do so. It might have been sufficient for the Arabs to be perceived as preparing to launch a full-scale attack on all fronts. This would have divided the IDF between the Egyptian, Jordanian and Syrian fronts, preventing a major Israeli attack on any of them due to lack of troops. It might have also weakened the IDF in some sectors to such a degree that its troops would not have been able to push back a determined Arab offensive. The proximity of some Israeli towns and kibbutz to the border on all fronts added to the pressure on the IDF to split its forces. In sharp contrast to an Arab assault in a low-intensity war, an Arab attack in a high-intensity war would have been much more powerful, and it might have carried enormous risks for Israel.

In the border wars an Israeli action against an Arab state was meant to deter the latter, and it also often diminished the will of other Arab states to assist the one that was confronted by Israel. In the 1967 showdown the swift Israeli victory on the Egyptian front had discouraged other Arab militaries from joining Egypt. Jordan and Syria did not commit their militaries to a serious offensive, as would expected from allies. The two Arab states opened fire, but, apart from few air bombardments and shelling on the Israeli rear, the Syrian and Jordanian fire was mostly around the border. So while a high-intensity war raged in Sinai, on the Jordanian and Syrian fronts there was mostly a border war.

A kind of low-intensity war between states that took place on the Jordanian front ended quite fast on the first day of the 1967 showdown. Israel restrained itself until its government decided that the Hashemite kingdom had gone too far within the vague rules of the border wars. The relatively massive Jordanian bombardment near the border, including Israel's capital city, together with capturing of a UN facility, was too much for Israel, which retaliated by commencing a high-intensity war; among other things, this development was meant to expand Israel's strategic depth, as indeed it did.

In the Golan, the Syrian defense depended on its forward line,[38] which was based on the assumption that the IDF only planned to attack Syrian posts near the border, as it did during the border wars in the previous years. Syria might have also relied on some concept of reciprocation, since during the 1967 showdown Syria did not try to invade deep into Israel and bombed Israeli villages just as it did in the border wars, at most initiating a few attacks on places close to the border. Israel, however, had a different idea, executed five days after the fighting had started. Israel's goals were much larger in comparison to those of the border wars. The border war in early June 1967 on the Syrian front was used, like the one on the Jordanian front, as a reason to start a major offensive—that is, a high-intensity war.

The Israeli offensive against Syria and the conquest of the Golan put an end to ongoing issues from the border wars: the "war over water" (a struggle for the control over the water sources of the Sea of Galilee) and Syrian fire on Israeli settlements in the Jordan valley. As long as Syria owned the Golan, it could have tried again to divert water from the Sea of Galilee and to shoot at the Israeli population. A full-scale Israeli attack in a high-intensity war in 1967 accomplished what many Israeli strikes had failed to do during the border wars on that front in the early and mid-1960s.

The PLO also played a role in the events leading to the 1967 show-down.[39] The low-intensity war between Israel and the PLO contributed to the tension that led to a high-intensity war in 1967. The PLO's plan was to turn this low-intensity war with Israel into a high-intensity war between Israel and Arab states, which should have ended in the destruction of Israel. The PLO got its wish, but not the results it was looking for, to put it mildly.

In the 1960s the United States considered Israel a potential ally.[40] However, being then focused on Vietnam,[41] this low-intensity war in Southeast Asia prevented the United States from increasing its effort to stabilize the Middle East and lower the chances of the 1967 high-intensity war happening.

## From Jordan to Lebanon: 1968–1982

After the 1967 showdown, Israeli soldiers were deployed on the west bank of the Suez Canal in tents and improvised rifle fits, which exposed them to dense artillery fire. It was two years before they had fortifications that gave them much better protection. This war of attrition between Israel and Egypt in the late 1960s included mutual exchange of artillery, sniper fire, commando raids, ambushes, and so forth.[42] Israel sometimes considered ways to avoid this kind of enduring and exhausting border war, which was more suitable to Egypt because of its superiority in numbers. Israel might have escalated the fight by launching a massive attack in which the IDF had an advantage, such as in maneuver warfare, but that would mean starting a high-intensity war. Ultimately Israel refrained from escalating the situation in spite of its frustration of the stalemate on the front line.

In 1970, after three long years of bitter struggle, Egypt failed to push Israel out of Sinai, or at least away from the Suez Canal, by relying on low-intensity war tactics. This drove Egypt to try a high-intensity war in 1973. Syria reached this decision too, after its border wars with Israel in the Golan in 1967–1973, although the Syrian operation was on a much smaller scale compared to the collision on the Egyptian front. However, following the 1973 high-intensity war, there were low-intensity wars on both the Egyptian and the Syrian fronts. It was a minor war of attrition in which the two sides tried to inflict casualties and harass the other for a few months, mostly by opening fire with artillery, small arms, and so forth.

In the late 1960s the IDF made an effort to stop the infiltrations, shooting and shelling from Jordan. Those assaults were aimed at civilian and military targets near the Dead Sea, in the Beit She'an valley and in the West Bank, which was, starting in 1967, under Israeli control.[43] Those low-intensity wars expressed the Palestinians' frustration as well as their determination to keep on fighting even after the Arab failure in the high-intensity war of 1967. Israel, standing firm against the Egyptian and Syrian militaries, did the same when facing the Palestinian assaults. Syria and Egypt, being busy on their own fronts, did not feel compelled to help the Palestinians by becoming involved in a high-intensity war, a collision they wished to avoid until 1973. When Egypt and Syria confronted Israel in 1973, the PLO ran a very limited low-intensity war on the border with Lebanon.

After the 1967 showdown, Iraqi troops, who came to assist Jordan during that high-intensity war, stayed in the Hashemite kingdom and undermined its internal stability.[44] They might have caused a low-intensity war between the local regime and the PLO when the latter thought it had an opportunity to take over the state. The Palestinians could have assumed that the Jordanian military would hesitate to confront them, since Jordanian troops were pinned down and neutralized due to the Iraqi presence in the kingdom. In 1970 the ongoing tension burst out when the PLO clashed with the Jordanian government, but it lost and had to flee to Lebanon, where it resumed its low-intensity war against Israel.

One of the Israeli goals in the 1982 war in Lebanon was to decimate the PLO's infrastructure of terror and guerrilla warfare.[45] Ariel Sharon, as Israel's defense minister, also wished to remold the Middle East by destroying the PLO, weakening Syria, making Lebanon an Israeli ally and boosting the position of the United States in the region at the expense of the Soviet Union.[46] Israel's chief of general staff in 1982, Refael Eitan, thought Israel should focus on the PLO as opposed to Syria.[47] Eventually Israel ran a high-intensity war against Syria in Lebanon and a hybrid war against the PLO, which was in 1982 a combination of a terror organization and a conventional military.

In 1982 the PLO was left by the Arabs to fight alone against Israel.[48] Syria likewise did not receive any help from Arab states reluctant to join the clash in the name of Arab solidarity. In the 1982 war Israel had a peace treaty with Egypt since 1979, but it was too soon to know after just three years if this agreement had survived a major collision between Israel and the Arabs. As it turned out, it did. Other Arab states, which did not have such official relationships with Israel, acted in the same way as Egypt. Furthermore, Palestinians who came under Israeli control in 1948 or 1967 did not start a rebellion against Israel while the latter confronted the main Palestinian organization. The Palestinians in the Gaza Strip, West Bank and the rest of Israel did not begin a low-intensity war during the hybrid war in Lebanon. They continued their policy of high-intensity wars, including during the first stage of the 1973 showdown, when Israel had its darkest hours since 1948.

## Comparing Israel's Allies from the 1956 and 1982 Wars

Israel had substantial difficulties in the wars before 1956 and 1982—that is, those in 1948–1949 and 1973. Therefore, in 1956 and 1982 Israel strove to demonstrate to the Arabs its military might and to achieve better results in comparison with previous rounds.

The 1956 and 1982 wars were similar in other ways as well. For example, they were both wars in which Israel planned in advance to team up with an ally. In 1956 Israel cooperated with France and Britain in the high-intensity war against Egypt. In 1982 Israel collaborated with Christians from Lebanon in a hybrid war against the PLO.

Israel and Jordan had clashed with each other before the 1956 war, and there was always the danger of Britain, Jordan's supporter, attacking Israel. Furthermore, owing to the defense treaty between Egypt and the Hashemite kingdom, an Israeli invasion into Sinai, according to its agreement with Britain, might have pushed Jordan to confront Israel.

Jordan might not have launched a full-scale attack against Israel, but, considering Israel's lack of strategic depth on the Jordanian front, even a relatively minor Jordanian assault there could start an Israeli offensive in the West Bank.

Ultimately Jordan kept out of the war, thus saving Britain from the complicated dilemma of whether to assist Jordan. In the 1982 war the PLO hoped that other forces in Lebanon—mostly the Shiites and Druze—would join it against the IDF. At that time, the Christians of Lebanon had endured several years of a brutal civil war against other Lebanese groups. Nevertheless, the unwillingness of the Shiites and Druze to resist the IDF spared the Christians of Lebanon an uneasy political and military situation—that is, fighting against other Arabs of their own country while collaborating with a non-Arab state: Israel.

In late October 1956 Ben-Gurion revealed his vision of reshaping the Middle East during the conference between Israel, France and Britain. One of his ideas, dividing Jordan between Israel and Iraq, was turned down by the representatives of France and Britain.[49] In 1982, what some saw as the old Israeli dream of using a war to remold the Middle East[50] was foiled, at least partly, because the Christians in Lebanon did not wish to participate in a grand plan of changing the region. Furthermore, for Britain in 1956 and for the Christians of Lebanon in 1982, Israel was the lesser of two evils. Britain and the Christians of Lebanon joined Israel to fight Arabs but only on an ad-hoc basis. Britain hoped to keep close ties with Arab states such as Jordan, and the Christians of Lebanon considered themselves part of the Arab world. Therefore Israel's aspiration in 1956 and 1982 to exploit those wars and their allies in order to achieve the goal of forming a new order in the Middle East was never realized.

In the conference before the 1956 war, France suggested that Israel start a war against Egypt in Sinai, thus giving France and Britain a reason to intervene by sending troops to the Suez Canal. Ben-Gurion was concerned that Israel would be left on its own, and he did not wish his country to look like the aggressor.[51] Before the war in 1982, the Christians from Lebanon also intended that Israel should start their joint venture alone. Israel opposed that idea but—as in 1956—finally accepted its allies' promise not to abandon it.

Before the 1956 war, France delivered to Israel dozens of jets and hundreds of military vehicles, including armor ones like tanks.[52] Starting in 1976, Israel started to give weapons and ammunition to the Christians in Lebanon.[53] In both cases the purpose of sending weapons was to assist a partner to fight its own battles. This emphasized the considerable gap between Israel's allies in 1956 and those of 1982. France and Britain were

European powers, while the Christians of Lebanon were just a relatively strong militia. Furthermore, in actual combat the French military in 1956 assisted Israel more than the Christians of Lebanon in 1982, who tended to make up excuses while avoiding contributing their share.

The difference between Israel's allies in 1956 and 1982 was seen also in the political arena. In the 1956 war France and Britain saved Israel from international isolation and condemnation in the United Nations. In the 1982 war Israel's hope that the Christians would create a pro-Israeli regime in Lebanon was not realized.

In 1982, after the massacre the Christians committed within Palestinian camps in Lebanon, Israel was held responsible for permitting them into those areas.[54] In the 1956 war Israel was blamed for collaborating with two colonial powers, Britain and France, against Arabs, including civilians.

In 1982 Israel sought to attack before the presidential elections in Lebanon took place in the summer of that year. The plan was to use the military campaign to put the Christian leader, Bachir Gemayel, into office.[55] The military operation was successful, but Gemayel was assassinated soon after he was elected.

## *The Low-Intensity Wars of 1987–1993 and 2000–2005*

Palestinians in the West Bank and the Gaza Strip chose to clash with Israel in two low-intensity wars, which took place in 1987–1993 and in 2000–2005. Those two collisions were not border wars, in contrast to almost all the low-intensity wars until 1987, since the fights in 1987–1993 and 2000–2005 occurred inside the geographic boundaries of Israel.

In the years 1948–1987 there were only two quite limited confrontations inside Israel. Following the 1967 showdown, there was an attempt by the PLO to stir up the West Bank, which caused minor outbreaks lasting a few months. In addition, in the Gaza Strip in the late 1960s and early 1970s there were often clashes between Israeli security forces and local terrorists.

A low-intensity war could have also occurred between Israel and those Palestinians who have been citizens of the state of Israel ever since 1948—Israeli Arabs. Yet, over the years, only a very tiny percentage of them have been involved in violent acts against their government. Those few insurgents did not manage to convince the rest of the Israeli Arabs to start a low-intensity war against their state, not even during wars between Israel and non-Israeli Arabs. Furthermore, in the 2006 war many Israeli Arabs were at risk just like most of the Israeli Jews, and sometimes even more, since most of rockets and missiles fired from Lebanon hit the north of Israel, where many Israeli Arabs live, while most of the Jews are concentrated in the center of the country. The 2006 hybrid war, when Israel was struck inside the country, and not only near the border, put many Israelis, Arabs or not, in the same boat. While many Israeli Arabs might have assumed that Israel is also—if not mostly—to blame for the 2006 collision, and that the IDF's attacks inside Lebanon—meant to protect Israeli Arabs too—only made things worse, they stuck to their traditional neutrality.

In the early 1990s Israel agreed to create the Palestinian Authority (PA) in order to prevent another low-intensity war like that of 1987–1993. If this kind of clash happened again, Israel wished to let the PA perform most, if not all, the dirty work. Such a move could also reduce the tension between Israel and Arab states and all that this tension entails regarding the probability of having another high-intensity war between them and Israel. The hope on the Israeli side was that the Arab world would be less concerned if an Arab government such as the PA suppressed its population than in the case of Israel doing that. Yet Israel took a risk by giving the PLO a political and military springboard against it, not only in a hybrid or low-intensity war but also in a high-intensity war, should the PA invite Arab militaries into its land. Furthermore, the PA was to a large extent involved in the collision of 2000–2005, which proved the failure of the Israeli concept about the original role of the PA. Fortunately, this low-intensity war did not cause a high-intensity war.

In the years 1987–1993 and 2000–2005, the Palestinians expected help from Arab states, be it only political and not military. There was less chance, then, of low-intensity wars inside Israel developing into high-intensity wars between Israel and Arab states.

In 1987–1993 and 2000–2005 Palestinians who fought Israel were not subordinated by Arab states or located there, in contrast to previous wars. This reduced the probability of entangling Arab states in a showdown with Israel. Arab regimes expressed their sympathy and support for the Palestinian cause, but after the collisions of 1987–1993 and 2000–2005 they started to view the Palestinian issue in general as more of an Israeli matter.

Nevertheless, Arab states could use the Palestinian cause to their advantage. For example, in 2002 Egypt could have dispatched forces into

Sinai, following Israel's penetrations into Palestinian "A" zones,[56] the heart of the PA. Such an Egyptian move would have violated the peace treaty with Israel and caused a severe crisis between them—maybe even a high-intensity war, if Egypt had insisted on keeping its troops in Sinai. Deploying Egyptian forces in the peninsula served not only the Palestinians but also—if not mostly—a vital Egyptian interest: changing the status of the peninsula as a demilitarized zone. Egypt got back all of Sinai in the early 1980s, but it had to tolerate constraints on the deployment of its soldiers in the peninsula. Israel could not have afforded to let the Egyptian military, in a position of force near its southern border, into the Negev. In spite of the peace treaty with Egypt, Israel learned from the 1973 war that it should plan its strategy based not on the expected intentions of the other side but on its actual actions.

A strong military presence of Egypt in northeast Sinai could have also prompted Palestinians, mostly those in the Gaza Strip, to provoke Israel, which would have raised the probability of a high-intensity war between Israel and Egypt. Therefore, in 2000–2005 Egypt chose to restrain itself from dispatching troops to Sinai. Egypt took only diplomatic steps, such as warning of an escalation, condemning Israeli actions against the Palestinians and calling back its ambassador back home. It also did not do its best, to put it mildly, to prevent the smuggling of weapons and other war material from Sinai into the Gaza Strip.

In 2000–2005 Bashar al Assad of Syria was not willing to jeopardize his state in a high-intensity war—certainly not for the PA. The Syrian leader did not assume that his country had an opportunity to retake the Golan in a high-intensity war while the IDF was occupied in the West Bank and the Gaza Strip. His father, Hafez al Assad, had thought the same in the former low-intensity war, that of 1987–1993, although he still had his Soviet patron for most of that time. (He could have perhaps concluded that this declining empire might not assist him in a high-intensity war, as it did in previous ones.)

During the first half of the low-intensity war of 1987–1993 Iraq could have started a high-intensity war against Israel. Iraq, following the end of its long showdown with Iran (1980–1988), had about two years until its invasion to Kuwait in August 1990, during which time it could have turned its enormous military with its battle-hardened troops against Israel, but it did not. Jordan, the buffer between Israel and Iraq, would have had to permit Iraqi units to enter its land, as was done in former high-intensity wars against Israel. Yet in the late 1980s the Hashemite kingdom had no interest in clashing with Israel, certainly not on its territory and at its expense, although strong Iraqi pressure, which could have included a clear threat, might have forced Jordan to yield to its much stronger neighbor to the east. Fortunately for both Israel and Jordan, Iraq decided to bully and then seize another Arab state that was even more defenseless than Jordan—Kuwait.

U.S. victories over Iraq impacted on Israel's low-intensity wars with the Palestinians. The defeat of Iraq in the high-intensity war of 1991 reduced the probability of an Arab coalition against Israel, which would have helped the Palestinians in their struggle at the time. A massive Iraqi deployment in Jordan would have boosted the morale of the Palestinians, particularly those in the nearby West Bank. The latter could have also received assistance from the Iraqi military. The blow Iraq absorbed in 1991 had a negative effect on the Palestinians that resembles, in a way, the ramifications they endured after Egypt's catastrophe in 1967. The beating of a major Arab state in a high-intensity war, as occurred in 1967 and 1991, was a setback for the Palestinians; yet they went on with their fight. The collapse of Iraq in 2003 not only completely removed any chance of Iraq contributing to a high-intensity war against Israel, thus assisting the Palestinians, but also deprived the latter of direct (although very limited) Iraqi support in their low-intensity war.[57]

It is possible that, following the Israeli withdrawal from Lebanon in May 2000, the Palestinians considered Israel weak, which encouraged them to confront it later that year, in September.[58] Therefore, it's possible that a low-intensity war in the West Bank and the Gaza Strip was ignited by the ending of a hybrid/low-intensity war in Lebanon. Yet the Palestinians did not respond like that to previous Israeli withdrawals, including those in Lebanon in the 1980s. Furthermore, Arab states did not see the Israeli retreat from Lebanon in 2000 as an opportunity to commence a high- or low-intensity war against Israel. Egypt and Jordan kept their peace treaty with Israel. Syria did not try to regain the Golan by force. In 2000 Israel's last high-intensity war had been 18 years in the past, and even that was quite limited and only against Syria. Furthermore, in 2000 Arab states remembered the 1973 showdown, when, in spite of their success, they lost land, men and almost the entire war. In 2000 they were well aware of IDF's combat potential, especially considering its ongoing military buildup.

The strength of the Israeli air force and armor deterred Arab militaries, but not the Palestinians, who knew that those two powerful corps would not play a decisive role in a low-intensity war against them. The Palestinians, who were much less strong than Syria, Egypt or Jordan, nevertheless assumed they could confront Israel. Indeed, they were right up to a certain point, since Israel seemed more vulnerable in hybrid/low-intensity warfare than in high-intensity wars. Of course, Israel could have easily crushed the Palestinians, but the ramifications for Israel, such as the political cost, might have been too much.

In a way, in 2000 the Palestinians implemented Egypt's strategy from the 1973 showdown to forego the hope of annihilating Israel in one giant strike. In 1973 the Egyptian goal was much more modest: to conduct a high-intensity war in order to create a political process in which Israel would return part of its land. This plan worked for Egypt, which eventually got back the Sinai. The Palestinians in 2000 adjusted this approach to their own circumstances. They believed that starting a confrontation—in their case, a low-intensity war—would lead to an agreement in which Israel would deliver the West Bank and the Gaza Strip to the Palestinians. This was in contrast to the era of high-intensity wars, when the Palestinians, together with Arab states like Egypt, strove to destroy Israel completely. For Egypt in the 1970s, as for the Palestinians in 2000, territorial compromise was acceptable. Nevertheless, the dream of all Arabs (particularly the Palestinians), to gain control of all the land of Israel, remained intact.

## Hamas as a Hybrid Rival

After the Hamas took over the Gaza Strip in 2007, Egypt's policy toward this group was ambivalent. Mubarak, on the one hand, refused to get into a conflict—let alone a high-intensity war—with Israel during its hybrid campaign against Hamas in December 2008 and January 2009. On the other hand, Mubarak did not prevent the smuggling of weapons from Sinai to the Gaza Strip.[59] This Egyptian tolerance enabled the Hamas to build its force in order to clash with Israel, which in turn could have entangled Egypt in the fight, including an escalation into a high-intensity war.

Starting in June 2012, Egypt had a new president, Mohammed Morsi, for about a year. This man has been identified with the movement he came from, the Muslim Brotherhood (MB), which has close ties with Hamas. Yet Egypt under Morsi also wished to stop another hybrid collision in the Gaza Strip between Israel and the Hamas for the same reason that Mubarak did: to avoid a high-intensity war with Israel. Although the MB has been hostile toward Israel in the past, in 2012 the movement had more urgent priorities: tightening its grip over Egypt and improving (or actually saving) its economy from disaster.

According to the 1979 peace treaty between Israel and Egypt, the Sinai Peninsula is subject to Egyptian rule; yet it must stay demilitarized. In August 2011 Israel approved an Egyptian request to send about a thousand troops and armor vehicles into Sinai to confront terror and guerrillas there.[60] Such permission to reinforce Egypt's presence in Sinai was rare. Usually Israel resisted any attempt to increase the number of Egyptian troops in Sinai. The Israeli dilemma was how to keep Sinai demilitarized as much as possible while assisting Egypt to win a low-intensity war in the peninsula against the terrorists. This was a delicate balance, because creeping Egyptian military in Sinai could lead to dangerous friction between Egypt and Israel, including the risk of a high-intensity war. The Egyptian forces did not need armor brigades for their confrontation with terrorists—only infantry ones with some armor vehicles and well-trained police officers.

Since 2013 Egypt made a serious attempt to cut off, or at least reduce, the smuggling of weapons from Sinai to the Gaza Strip. While it was difficult for Egypt to handle the chaos that spread in the vast area of Sinai—some 60,000 square kilometers—its border with the Gaza Strip is only 14 kilometers long. This made it easier to limit smuggling, thus reducing the possibility of a hybrid war between Israel and the Hamas, with all its known ramifications for Israel and Egypt.

## Hezbollah as a Hybrid Rival

In 1982 the hybrid war between Israel and the PLO and the high-intensity war between Israel and Syria in Lebanon led to a long low-intensity war between Israel and the Hezbollah in that country. In the 1990s this ongoing fight gradually turned more and more into a hybrid war, following the upgrading of the Hezbollah to a military outfit.

In the mid- to late 1990s Israel and Syria held diplomatic negotiations concerning the Golan. Those talks did not prevent Syria from supporting Hezbollah's military activity directed against the IDF in south Lebanon, and in fact pushed such support. Syria took this risk because it sought to drive Israel out of Lebanon while using the campaign there to put pressure on Israel to give up the Golan in return for Syria ceasing to assist the Hezbollah.

Syria played a dangerous proxy game, which might have dragged itself and Israel, against the will of one or both of them, into a high-intensity war with each other.

In a way, Syria won the battle in Lebanon, since Israel, of its own volition, withdrew from there in May 2000. However, Syria did not get the Golan back, and its own army was also forced to leave Lebanon five years later, in 2005, partly because of the momentum caused by the retreat of Israel. Thus, ironically, the hybrid activity of the Hezbollah did not lead to the repossession of the Golan by Syria, and Bashar al Assad had to give up another area that he believed belongs to him—namely, Lebanon.

Following the 2006 war in Lebanon, Assad, who was impressed by the performance of the Hezbollah, threatened Israel with a confrontation if no negotiations on returning the Golan to Syria were pursued. As a result, the two states prepared for battle. Still, Assad, like his father before him, avoided any clash in the Golan, not even with proxies such as Palestinian refugees in Syria. The Syrian regime, much as it wished to regain the Golan, feared the ramifications of a collision on that front, even a limited one. Igniting a clash in nearby Lebanon was less dangerous than confronting Israel from Syria itself. This explains why ever since 1974 and until 2011 there was a permanent quiet in the Golan, the longest period of calm on one front in Israel's history. It was a kind of an unofficial peace based on deterrence, not on a treaty similar to that with Jordan and Egypt. Furthermore, Egypt did not completely stop the smuggling of weapons and manpower to guerrilla and terror groups in the Gaza Strip that conducted hybrid or low-intensity war against Israel.

However, Israel paid in blood for its presence in the Golan through the hybrid and low-intensity war against the Hezbollah, which had Syria's support. In a way, the collision in Lebanon replaced a low-intensity, hybrid and even high-intensity war in the Golan, which—as far as Syria was concerned—had the same aim: returning the Golan to its hands. In addition, there was much less chance that Egypt, let alone the Hashemite kingdom, would commence a high-intensity war against Israel, while with Syria this probability was not that small. Contrary to Egypt and Jordan, Syria had deployed several divisions near the Israeli border, in the Golan, and they were not there for defensive purposes only.

Considering the mutual suspicion and the number of troops near the border in the Golan, a third Golan war (after those in 1967 and 1973) might have broken out. One scenario indicated a possibility of one side assuming, if only by mistake, that its rival was about to attack, which could have pushed that side to launch a preemptive strike.

The Hezbollah supported the Palestinians in their fight in 2000–2005 by delivering weapons, money, knowledge of how to make deadly IEDs, and so forth.[61] As Syria did in Lebanon with the Hezbollah, the latter ran a proxy war, a low-intensity one, for the same reason: to evade a direct collision with Israel. Syria wished to avoid a high-intensity war, while for the Hezbollah it was the risk of a hybrid war. In fact, the 2006 war in Lebanon started after the Hezbollah conducted one incursion on the border with Israel, pushing the latter to attack, and not because of all the provocations of the Hezbollah against Israel, helping the Palestinians, and so forth.

# 4

# The IDF in Fighting
# Low- and High-Intensity Wars

The IDF had to be ready to deal with both low- and high-intensity wars. The military similarity between Israel's high- and low-intensity wars lay mostly in testing troops under fire, in both defense and offense, while using weapons ranging from small arms to tanks, artillery and planes. The difference was often in the scope of the operations, order of battle, and so forth.

## The IDF in the 1948–1949 War

During the British Mandate, Jewish low-intensity warfare demoralized the British security forces and eventually, according to Daniel Byman, played a "major rule in convincing the British to leave."[1] The fight between Israel and Arab militaries broke out subsequent to the British withdrawal. During the first stage of that showdown, there was no IDF. Israel had only a bunch of armed groups experienced in low-intensity warfare against the British military and police, and Arab outfits. Even after the IDF was formed, it was, at least in its beginning, more of a semi-conventional organization than a modern military. It possessed hardly any planes, tanks, artillery, and so forth. Many of its troops were thrown into combat with little, if any, training, often with almost no equipment or standard uniforms. There were also youth and women who were involved in the battles and in towns and convoys that were attacked. The IDF was a very fresh, poorly armed and barely trained ragtag military.

During the 1948–1949 war this newly established army adjusted to the requirements of high-intensity warfare by steadily building and upgrading its strength.[2] Until that point, from the Arab perspective, it could have seemed that their enemy was a kind of guerrilla military, or at most a hybrid one, not a force to be reckoned with.

However, the Arabs failed to turn their early advantages into victory in combat, while Israel, having started a high-intensity war with an improvised military that was better suited to low-intensity war, gradually overcame the Arab militaries. During the showdown the IDF conducted large operations, which were typical of high-intensity wars, along with much smaller actions like raids, more in keeping with low/hybrid collisions.

## Coping with Palestinian Infiltrations and a Possible Arab Invasion in the 1950s

In the 1950s the IDF had to be ready to foil a massive Arab attack.[3] The IDF also faced Palestinian infiltrations, although members of its planning branch claimed the military should avoid dealing with such tasks.[4]

Israeli artillery on the Egyptian front, outside Beer Sheba during the 1948–1949 war. (Photographer: Hugo Mendelson; Source: Israel's Government Press Office)

This group preferred that the IDF stay out of (or at least reduce its involvement in) the border wars, in order to focus on preparing for a high-intensity war. Indeed, not only was the latter challenge more critical, but it could only be handled by the military, whereas the infiltrations could be taken care of by the police. However, while the killing, harming and terrorizing of innocent Israeli citizens by Palestinians was equivalent to a crime wave, the police, busy with regular crimes, lacked the equipment, training and manpower to halt the infiltrations all on their own. The IDF, with all its difficulties in the early 1950s, including budget cuts, was still much more powerful than the police force.

In order to invade Israel in a high-intensity war, an Arab military needed roads for its various vehicles. Some fronts were easier to travel through, such as the open desert of the Negev. Furthermore, there were areas on the border with Egypt that had no natural or artificial obstacles to delay an Egyptian force. All those advantages in favor of the enemy presented the IDF with a difficult challenge, which was also manifested in the border wars (albeit in a minor form)—Palestinians could sneak in by foot from nearly every sector.

Israel's lack of strategic depth in the West Bank meant that an Arab offensive there could reach Israeli cities before the IDF managed to halt them. Therefore, when a warning about an Arab attack came up, the IDF immediately had to initiate a high-intensity war—that is, launch a preemptive strike or a preemptive war.

During the border wars, Palestinians took advantage of the short distance between the border and Israeli towns and villages. Israel absorbed those tiny infiltrations without striking first, as was needed in a high-intensity war, despite its preference of stopping in advance any assault on its civilians. Such a defensive move was not always an option for Israel, because many times in the border wars there were no specific alerts about Palestinian incursions.

During the 1950s David Ben-Gurion suggested occupying areas in the West Bank in order to cope with Palestinian infiltrations from that area.[5] This meant a high-intensity war with Jordan, the ruler of the West Bank at the time. Ben-Gurion's idea was probably based on the fact that Israel already had areas populated with Palestinians, more than 150,000 of them, and they did not turn into insurgents. In the 1950s, if Israel had conquered the West Bank, the Palestinians residing there might have avoided expressing hostility toward Israel. However, some of them might have only pretended to accept the Israeli government while awaiting an opportunity to strike.

Another possible outcome of a high-intensity war in the 1950s whereby Palestinians in the West Bank would be pushed to relocate to the rest of Jordan could have been a renewed fight on their part from their new home (as occurred when the Palestinians were forced to abandon their former territory during the 1948–1949 war).

## Raids in Comparison to Major Attacks in the 1950s

The origins of the IDF's offensive approach were rooted in the late 1930s, when Jews started to go out of their villages to confront Arabs in the open field and in Arab towns.[6] The IDF upgraded this strategy during the high-intensity war of 1948–1949.

In the 1950s, as in the 1930s and 1940s, Jewish civilians, now citizens of Israel, were attacked by Arabs, now defined as Palestinians. The response was similar: transfer the fighting as far away as possible from Israeli cities and settlements. This was also the concept to implement in the case of another high-intensity war.

The IDF attacked military camps in both low- and high-intensity wars. The challenge for the IDF was greater in a high-intensity war, following the scope of its operations and their aims. A raid on a police station during the border wars was not like conquering a major compound in a high-intensity war. However, both those missions had to end quickly.

In the border wars, Israel's concerns were local Arab military reinforcements pouring in, while in high-intensity wars the worry was that other Arab states and even foreign powers could use military or political means to intervene.

In the high-intensity war of 1948–1949, the IDF had sometimes deployed several brigades on one front, as it did in late October 1948 against Egypt.[7] In contrast, only one brigade participated in a raid in early November 1955,[8] and even this was a rare example. In all the other raids during the border wars of the 1950s, the scale of the Israeli forces was much lower—a battalion, company, and so forth. Yet, as in a high-intensity war, during the border wars the IDF concentrated its efforts in a certain sector, by sending elite troops, for example, which improved the chances of success.

In the 1948–1949 war the IDF beat the Arab Liberation Army, but it did not annihilate any of the Arab regular militaries. At most, the IDF destroyed parts of them stationed in posts. In the border wars, the IDF often wiped out its foes, gathered in much smaller quantities than in high-intensity wars. They were mostly garrisons of an Arab military camp or a police station.

The Israeli paratroops, about a battalion strong in the early 1950s, carried out many of the raids during the border wars, which stretched over several years. The time between each of the raids and their quite limited scale allowed the IDF to dispatch immediate reinforcements to the paratroops' aid if they ran into trouble. In high-intensity wars the IDF had to spread its resources over several fronts, but it had far more manpower at its disposal because the reserves were mobilized, while border wars were based on a scanty regular army.

In the raid on a police station in Kalkilia on 10 October 1956, the IDF lost 18 men. It was obvious at the time that Israel could not have continued with those kinds of strikes.[9] This was too high a price to pay, particularly since the IDF brass sought to minimize its casualties as it aimed for in high-intensity wars.

Dayan, as the chief of general staff, claimed that the IDF should be ready to launch a raid in less than 24 hours, even without proper preparations and while risking substantial casualties. According to Dayan, if the IDF was not able to execute a rapid-pace operation during border wars, its forces could hardly be expected to do so in a high-intensity war.[10] Indeed, particularly during an Arab surprise attack on several fronts, the IDF had to deploy its troops in a matter of hours, not days.

However, it was difficult to reach conclusions in this matter based on information gleaned from raids in the border wars (which required only a few hundred troops), since gathering tens of thousands of soldiers was necessary in a high-intensity war, and a massive mobilization automatically required much more time and resources.

Furthermore, it should be emphasized that in the 1950s the IDF was a new military that had no combat experience in amassing reserves in a high-intensity war. The call to arms was a process that included unknown risks that were not part of the border wars, such as Arab bombardments on the Israeli rear that could have disrupted the gathering of the reserves.

The IDF ran vast scale exercises in the 1950s.[11] Such drills were not as realistic as actual combat in the border wars, but they were the most effective way—if not the only way—to train troops in basic aspects of high-intensity warfare. For example, the penetration of an armor brigade dozens of kilometers deep into "enemy" land was one of the maneuvers that could not have been examined in the border wars.

# The Crisis of the Infantry in the Border Wars, 1949–1956

The infantry was the main corps in the IDF in the early and mid-1950s,[12] resulting from its center role in both the 1948–1949 war[13]—the biggest combat test of the IDF until that time—and the low-intensity wars before 1948 (that is, the underground activity against the British Mandate). This stormy history made the infantry the most veteran corps in the Israeli military and, as a result, the most powerful one.

In 1953 there were problems when infantry troops went on raids. Sometimes regular units failed to penetrate Arab villages.[14] Although the IDF had gained experience in this type of combat in the 1948–1949 war, after a few years its soldiers did not always manage to accomplish similar tasks in the border wars. This raised deep concerns, and an obvious question: If the infantry was pushed back in an Arab village, how could its men storm a well-fortified post in a high-intensity war? Furthermore, in the border wars, during almost any raid, Israeli troops could withdraw if they ran into trouble, because it was just another skirmish in a long struggle that lasted for years, and the IDF could try attacking the same site again at another time. However, in a high-intensity war, particularly since the IDF strove to win fast, the objective had to be taken as soon as possible. In the border wars there was usually only one target, while in a high-intensity war there was an effort to conquer several posts in the same area, in order to secure an entire sector. The results of battles in each destination could have a major impact on the clashes in other places.

In a border war the purpose was to raid one location and then retreat. In a high-intensity war the troops had to hold the point while expecting an Arab counterattack.

The IDF could tolerate embarrassing failures in border wars if that meant avoiding unnecessary casualties that would be a blow to the IDF and an achievement for the Arabs. The underlying assumption was that in a high-intensity war the performance of the Israeli troops would be much better, owing to the success of their commanders convincing them of their duty to accomplish their mission at any cost. Yet this approach was too much of a gamble for the IDF, considering the difficulty of adjusting the troops so quickly to a different pattern of action during an ongoing high-intensity war and the negative ramifications of a defeat for Israel.

Regular troops had a key role in both preparing the IDF for a high-intensity war and actual fighting, such as holding the lines until the reserves joined them.[15] Since the IDF relied on the infantry, the regulars of that corps had special importance in training the military for high-intensity war and in conducting it. The regular troops were also those who carried out most of the raids. The blunders in those clashes were therefore associated with the regular soldiers. Due to the importance of the latter, the crisis of the border wars indicated that the IDF was in a pretty miserable condition.

Following an immediate and pressing demand for reform, several actions were taken, one of which was the creation of a special unit, the "101."[16] This elite infantry formation of selected warriors served as a role model by demonstrating to the rest of the IDF, and particularly the infantry, how to find and confront the enemy. The missions of the "101," such as tactical reconnaissance, were relatively minor tasks in a high-intensity war, but the willingness of its men to fulfill their assigned duties was their most valuable asset. This concept, emphasizing morale and discipline, was supposed to help the rehabilitation of the infantry and its preparedness for border wars as well as for high-intensity wars.

**Israeli troops in armored vehicles on the Egyptian front during the 1948–1949 war. (Source: Israel's Government Press Office)**

# The Armor and the Infantry, 1949–1956

Following Western policy toward the Arabs and Israel in the early 1950s, the latter had was constrained in acquiring many tanks, fighters, bombers, and so on, for its armor and air force. The IDF had a limited number of those weapon systems, which made it more difficult to win a high-intensity war, although the IDF was based on the infantry and not on the armor and air force. This problem had no influence on the border wars, since tanks and planes were almost never used during raids, let alone in large numbers.

In the early 1950s the armor was considered by many in the IDF to be a corps that plays a secondary role.[17] In March 1952 the tanks had no ammunition for more than one day of combat.[18] It is no wonder its officers demanded that the status of the armor corps be strengthened.[19] The IDF at the time had 11 infantry brigades and only two armor brigades.[20] The armor hardly played any part in the border wars of the 1950s, if only because involving it could have caused an unnecessary escalation. Thus, the armor did not get a chance to prove its capability in the border wars, which reflected on its rather poor status in the IDF, including its presumed importance in a high-intensity war.

Israeli armor officers called to reshape the military into an armor-based one, but they were too few and did not carry much weight.[21] For them, it was not sufficient just to gain more tanks. The IDF also had to decide whether to continue with its current structure—a military based on infantry—or put the armor in a leading role. Calculations of cost and maintenance, combined with mistrust among the IDF's top brass regarding the tank's potential, slowed down the development of the armor in the IDF. The common belief within the IDF was that the infantry was the dominant corps, while the tank only assisted. The IDF ignored the possibility that if this concept failed in a high-intensity war, contrary to what happened in the border wars, the organization might not have enough time to correct its mistakes.

Israel's armor corps obtained the AMX-13, a tank that fit perfectly with the ideas of Dayan, who claimed that the tank's roles were patrol and combat support for the infantry.[22] Almost half of the tanks in the IDF (180 pieces) were AMX-13s.[23] It was a light tank, made for reconnaissance and not intended to encounter other tanks. This weapon system might have been too vulnerable in a high-intensity war. The weaknesses of many of Israel's tanks should have made the Israeli military leadership rethink about initiating a high-intensity war. Since the IDF counted on the infantry, the armor issue did not seem all that vital, but the reality of the battlefield of a high-intensity war was about to prove otherwise.

## The 1956 War—Operational Aspects

In the 1956 war Israel's 1st Infantry Brigade captured the posts in the sector of Rafih[24] in the northeast part of Sinai. The experience that the brigade gained from conquering strongholds in the border wars might have helped them in occupying the fortifications at Rafih. Like many raids during the border wars, the attack in Rafih in the 1956 war was conducted at night, which made it easier to surprise the defenders. In contrast to the border wars, the attack in the 1956 war included tanks from the 27th Armor Brigade, which supported the offensive in Rafih; afterward, they stormed deep into Sinai.

During the border wars, Arab militaries gradually learned how to better respond to Israeli raids.[25] The IDF also had to take into consideration that Arab militaries might improve their capabilities in fighting a high-intensity war. Yet in the 1956 war, which lasted a week (much shorter than the border wars), Egypt's military did not adapt effectively or quickly to Israeli warfare. Some Israeli methods were known to Egypt from the border wars, but, due to the limited nature of that low-intensity war, many Israeli combat patterns were revealed only in a high-intensity war, such as a rapid advance, deep penetrations, and so forth.

In the 1956 war the 10th Infantry Brigade assaulted Um Katef, a key compound in northeast Sinai. Its troops were pushed back because they were not determined enough.[26] This outcome indicated that the lessons from the border wars, following the activity of unit "101"—accomplishing the mission in spite of the danger—were not implemented. The result of the failure of the 10th Infantry Brigade was a delay in the Israeli schedule. As sometimes happened during the border wars, when the IDF ran into trouble, reinforcements were sent, and in this case the 37th Brigade was thrown into the battle of Um Katef.

Israeli soldiers gained confidence after the 1956 war[27] in spite of problems that occurred (such as the events at Um Katef). The IDF also obtained victories in the border wars, although only a relative handful of troops participated in them. The scale of the operations and the order of battle in the 1956 war seemed to have a greater impact on the average Israeli soldier in boosting morale. In addition, during the 1948–1949 war Israeli troops fought desperately to hold back a massive Arab offensive. That showdown was a terrible experience for Israel. Although the border clashes in the 1950s spread anxiety within the Israeli population, the greatest fear was another high-intensity war. This is why the Israeli troops in 1956 were willing to risk their lives in order to stop in advance any clear and present danger, especially a threat similar to the one that occurred in 1948–1949. This determination was reinforced after Egypt blocked the Tiran Straits and signed a huge arms deal, which many Israelis saw as a demonstration of Egypt's aggressiveness.

On the first day of the 1956 war, Israel, as part of its alliance with France and Britain, sent the 890th Paratroop Battalion to land deep inside Sinai. After one day the 202nd Paratroop Brigade reached its isolated battalion; yet they did not play a key role in the war.[28] In spite of their vast experience from the border wars, during the rest of the high-intensity war of 1956, the paratroopers were not dispatched to the main sectors—just to relatively minor ones, such as capturing the small and remote airport in A-Tor.

In the 1956 war the IDF found weak spots in its performance, such as in the command and control of a division,[29] which is quite an important factor in a high-intensity war. It is possible that the attention the IDF gave to the border wars was at the expense of investing time and resources in preparing for a high-intensity war, such as training its senior officers in running a division. The border wars were the most immediate challenge in the early and mid-1950s, although not necessarily the most important one, considering the ramifications of losing a high-intensity war due to unpreparedness of the IDF. Although the IDF won the 1956 war, it could not ignore its blunders. The next round could be much tougher. Lack of readiness in controlling a division, for example, carried grave danger for the IDF.

The IAF had in the 1956 war only 16 operational Mystere 4A jets,[30] its most advanced fighter jets at that time. They were too few in number to protect Israel and its troops in a high-intensity war. Without the help of Britain and France, which destroyed the Egyptian air force, Israeli cities and ground units would have been quite exposed. In the border wars this fact did not stand out, because air forces from both sides were not involved in the squabble. In a high-intensity war, however, air power played an essential role. It was another example of the limitations of the border wars as far as testing and preparing the IDF for a high-intensity war.

Following the alliance with France and Britain in the 1956 war, the IAF was not allowed to cross the Suez Canal—that is, to fly from Sinai

**Israeli paratroopers after they landed at the Mitla Pass during the 1956 war. (Photographer: Avraham Vered; Source: Israel's Government Press Office)**

into the rest of Egypt.[31] This meant that the IAF could not attack Egypt's strategic and military assets, since most of these were outside the peninsula. In the border wars the IAF did not see much action, let alone bombing remote objectives in Egypt. By choosing to restrain its air operations in both low- and high-intensity wars, Israel left targets deep inside Egypt unharmed.

In the 1956 war Dayan estimated that the nature of the Israeli maneuver would confuse the Egyptian senior command, who might hesitate to send planes to battle.[32] This Israeli thinking may have had its origins in the border wars—a successful move on the ground would astonish and even paralyze the foe. Eventually, however, Egyptian fighters were scrambled into Sinai in the 1956 war.

In the 1956 war an Egyptian destroyer was captured by the IDF, but only after the ship approached and shelled Israel's naval port in Haifa.[33] For Israel, this was clearly a different challenge in comparison to the infiltrations during the border wars. The IDF failed to prevent this naval incursion, and it responded only after the attack, as in many penetrations in the border wars. The rest of the Egyptian fleet did not attack Israeli objectives, possibly due to the presence of the armada of France and Britain in the Mediterranean. Egypt's navy stayed out of the fight with Israel, as it did during the border wars.

## 1957–1967: The Rise of the Armor and the Air Force

In the years 1957–1967 the IDF's buildup was based on the armor, following the rise of this corps after the 1956 war. The IDF realized that its current structure, which relied on the infantry, was questionable, if not utterly wrong.[34] What was sufficient for the border wars in the 1950s was not satisfactory in a high-intensity war. The armor had much more protection, firepower and speed than the infantry. These advantages were demonstrated, for example, when the withdrawal of Egyptian units from the peninsula, in the 1956 war, turned very soon into a total collapse, best exploited by the Israeli armor, which was faster than the infantry. The IDF recognized that the armor, not the infantry, should carry the main burden of winning the next high-intensity war.

The infantry continued to play a role in the low-intensity wars of the 1960s. One of the known battles took place when Israeli infantry raided a Syrian camp at Nokive in the Golan on the night of 16/17 March 1962. As in the 1950s, those operations helped the IDF learn how to overcome a fortified post, which was useful in case of a large offensive on the Golan in a high-intensity war.

In the mid-1960s Israel and Syria collided with each other in the "war over the water," a fight for the control over water sources of the Sea of Galilee. The Israeli armor used the skirmishes to get better prepared for a high-intensity war. The effort paid off in the 1967 showdown.[35] The border wars with Syria had contributed to the skills of Israeli tank crews, mostly in confronting enemy tanks. Although only a few armor troops took part in those skirmishes, the lessons, such as ways to hit distant targets, were shared with the rest of the armor corps. Israeli tanks also participated in a raid on the West Bank in November 1966. Such actions involving armor as part of the border wars were very rare, in order to enable the armor to focus on exercising for a high-intensity war.

"Tamar" was a vast drill, on the level of a division, executed in late November 1965. It examined various combat scenarios, such as tanks versus tanks, tanks against armored infantry and combined assaults of tanks, infantry, artillery, military engineers, and so forth. Another goal was to check and inspect how to deploy and maneuver with large numbers of tanks.[36] Such training was essential in preparing for a high-intensity war, and it could not have been tested in the border wars because of the order of battle of the exercise and its aims.

The IAF, as the armor, became one of the IDF's major corps after the 1956 war.[37] In the "war over the water" the IAF gained valuable knowledge in conducting bombardments and interceptions, planning under pressure of time, and running effective briefing.[38] In an aerial battle on 7 April 1967 the IAF shot down six Syrian fighters, all of which were Mig 21s.[39] In the 1967 war the IAF also intercepted and destroyed 37 Egyptian and Syrian Mig 21s.[40] Studying the weakness of the Mig 21 was one example of how the experience and skills of the IAF—developed and tested in combat during the low-intensity wars on the Syrian front—helped in a high-intensity war on two fronts.

On 4 June 1967, the IAF launched Operation Moked, which neutralized the Egyptian, Syrian and Jordanian air forces in a matter of hours, mostly by destroying their planes on the ground. This well-organized offensive was executed after years of training in navigation, gathering intelligence, producing bombs designed to tear up runways, and so forth.[41] Those steps were not connected to the border wars, and the IAF had to build and examine them in specific exercises and tests.

## The 1967 War

In the 1967 showdown, Israel had 1,000 medium tanks and 286 combat aircraft. Egypt alone had 1,300 medium tanks and 431 combat aircraft.[42] The balance of power tended even more in favor of the Arabs when Jordanian and Syrian forces were included. While in the border wars each side deployed only a tiny percent of its military forces, in a high-intensity war their order of battle was obviously much bigger. Nevertheless, as in major skirmishes during the border wars, in the 1967 showdown the main burden fell again on elite units such as Israel's 7th Armor Brigade, which broke through the massive fortifications of the Rafih sector.

On 4 June 1967, Israel's first strike was launched against Arab air forces that had years to prepare for such a blow.[43] As in Israeli raids on posts during the low-intensity wars, the element of surprise was critical. The aim was also similar: to destroy enemy forces, not conquer Arab bases. In 1967 the IAF annihilated ground targets, mostly planes, without seizing the airfields.

The invasion of the Israeli 31st Division into Sinai was a bold and a shocking move,[44] proving that even a formation of this size could infiltrate

**Israeli tank crews rush to their tanks on the Egyptian front during the 1967 war. (Photographer: Cohen Fritz Cohen; Source: Israel's Government Press Office)**

and sneak behind enemy lines, just as small commando units did in the low-intensity wars. However, even though the peninsula was a vast and open area (a fact that helped the 31st Division penetrate into it), the maneuver could have been spotted, because of Egypt's massive deployment in the north of Sinai, which expected an Israeli offensive.

In the 1967 showdown Israeli soldiers demonstrated high morale[45] because they felt their country faced an immediate threat. The collisions in the low-intensity wars in the early and mid-1960s did not contribute much to the willingness of Israeli troops to jeopardize their lives in the 1967 war. In contrast to clashes in the border wars of the 1950s, those in the 1960s, like the "war over water" on the Syrian front, had less of an effect as a model for fighting spirit. Most of the skirmishes with Syria were exchanges of fire, not as heroic as storming posts in the 1950s. In the 1960s (at least until 1967), there were also raids on Arab sites such as Syrian posts, but only few—much less than in the 1950s. In addition, the "war over water" was about preventing a problem in the future, while the raids in the 1950s followed assaults on Israeli citizens and were more inspirational events for the IDF. After all, the "war over water" regarded disrupting the water supply, which was a challenge but not as intimidating as attacks on Israeli citizens.

## Low-Intensity Wars on Three Fronts, 1967–1973

In spite of the clear Arab defeat in the 1967 showdown, the capital cities of Egypt, Syria and Jordan (and most of their infrastructure and populations) stayed intact, as in the border wars. The Arabs (mainly Egypt and Jordan) lost lands, but the strategic balance remained the same— asymmetry in favor of the Arabs in the population size, territory and natural resources. No wonder Arab states refused to recognize or even negotiate with Israel. They kept their basic strength, although they understood that in the near future it would be too difficult to challenge the IDF again in a high-intensity war. They preferred to rely on tactics based on low-intensity wars. The IDF continued to invest much of its resources in getting ready for another high-intensity war, but at the same time its troops had to deal with border wars on three fronts.

**Israeli artillery in Sinai during the 1967 war. (Photographer: Micha Han; Source: Israel's Government Press Office)**

In the late 1960s the IDF clashed on the Jordanian front with both the PLO and the Jordanian military. On 21 March 1968 the IDF conducted a vast operation inside the Hashemite kingdom, near and in the town of Karameh. This fight became complicated, mostly because Israeli units ran into killing fields in rugged terrains, where Israeli forces were exposed to effective Jordanian fire. 33 Israeli soldiers and 84 Jordanian troops were killed. Prior to this battle, some in the IDF had assumed it would be another success, as in the 1967 showdown[46] that had occurred less than a year before. Many of the Israeli troops who participated in the 1968 raid were from the 7th and 35th Brigades, which fought in the 1967 war on the Egyptian front, where they had their share of tough battles. In spite of that, some of them, like others in the IDF, apparently believed that their swift victory in the high-intensity war of 1967 meant they could easily overcome any Jordanian resistance in a low-intensity war. This hubris cost them dearly, in spite of their achievements in the raid.

In the Golan, as a result of border wars with Syria in the years before the 1973 war, Israeli armor units were accustomed to certain kinds of clashes.[47] Those limited collisions contributed to the techno-tactical skills of the Israeli troops, which helped them in the 1973 showdown. However, the operational conditions of the 1973 high-intensity war were much different from the exchange of fire in the border wars. The Israeli armor crews had to adjust immediately to a much bigger campaign—a massive Syrian invasion. Not all the armor commanders were aware of that significant change, because they were not properly briefed before the battle started. This was one of the key reasons why the IDF came close to losing the campaign in the Golan.

In the war of attrition that took place during 1967–1970 on the Egyptian front, the valuable experience the IAF gained in launching bombardments at night while dealing with surface-to-air missiles helped its air crews to be better prepared for the 1973 showdown.[48] However, the IAF suffered heavy casualties, and 102 of its planes were lost.[49] One of the reasons was that in the low-intensity war of 1967–1970 the IAF usually had time to regroup and re-engage the enemy on another day. By contrast, in the high-intensity war of 1973, particularly during the first days, when the Arabs advanced toward the Israeli lines, there was pressure on the IAF to bomb them immediately, sometimes regardless of the cost.

In the war of attrition in 1967–1970 the IAF bombed the Egyptian rear as part of the fight for the control of the Suez Canal.[50] In the 1973 showdown Israel conducted a strategic bombardment in the Syrian rear. As in the low-intensity war on the Egyptian front in 1967–1970, the bombardment on the Syrian front in the 1973 high-intensity war had to do with the fight on the front line. In both confrontations Israel's primary goal was to coerce an Arab state into agreeing to a cease-fire. Israel was aware that an Egyptian or Syrian retribution would not be easy to carry out. Those two Arab states knew it would be very hard for their bombers to reach a major Israeli city, since the Israeli fighters could have intercepted most, if not all, of the Arab planes. Nevertheless, in 1967–1970 and 1973 Israel was careful not to intensify its air offensive in order to avoid an intervention on the part of the Soviet Union—the patron of Egypt in 1967–1970 and of Syria in 1973.

Before the 1973 war, the IDF planned to send a division across the Suez Bay, with the intention of surprising the Egyptian military from its rear. This attack was linked to Operation Raviv, a daring raid that was carried out in the war of attrition on 9 September 1969. A relatively small Israeli force, about nine armor vehicles, had stormed the west coast of the Suez Bay and destroyed military facilities, equipment and vehicles. This limited operation from a low-intensity war might have inspired the IDF to initiate a similar move, only much bigger, in a high-intensity war, although in the end it was not executed.

**Israeli military helicopter during the raid on the island of Shadwan during the war of attrition. (Photographer: Moshe Milner; Source: Israel's Government Press Office)**

In the 1973 showdown, the IDF made one major vertical flanking by air: conquering the Syrian camp in Mount Harmon.[51] The IDF gained valuable experience in various airborne operations during the war of attrition of 1967–1970. Some of them were conducted deep inside Arab territory. These lessons included understanding how to coordinate between the helicopter pilots and the infantry troops who landed on the ground.[52] In contrast, Israeli airborne actions were few in number during the 1973 showdown, and most of them were relatively close to the front line. Although the 1973 high-intensity war lasted less than a month, and not years, as had the low-intensity wars, the vast knowledge the IDF possessed in this field and the three paratrooper brigades that were mobilized in 1973 should have been better used during the confrontation.

In Sinai the IDF had infrastructure consisting of communication, command and intelligence centers, and airfields.

During the war of attrition the IDF built near the Suez Canal a series of outposts, which gave Israeli soldiers reasonable cover from shelling, sniping, raids, and so on. Coping with a large Egyptian offensive required the upgrading of those forts with killing zones full of obstacles, anti-personnel minefields, and so forth. This was not done. The outposts were

**Syrian tanks in the Golan Heights during the 1973 war. (Photographer: Eitan Haris; Source: Israel's Government Press Office)**

also too few to be able to dominate every spot on the Suez Canal with effective crossfire.[53] There was a fundamental dispute inside the IDF regarding whether those little camps were also supposed to serve as a defensive line in a high-intensity war.[54] In the 1973 showdown, all those sites were captured by the Egyptians, except for one. These outposts— strong enough to push back Egyptian assaults in the low-intensity war— failed in a high-intensity war. The reasons for that failure were the overwhelming Egyptian force, inadequate preparations and lack of sufficient support from Israeli armor and air force.[55] The IDF had therefore what was at best a delaying action line, not one that could have stopped a full-scale offensive.

The ultimate objective of the IDF has always been to protect the Israeli people. In 1967–1973 its troops in Sinai, and especially those close to the border with Egypt, in the Suez Canal region, were more than 200 kilometers from the nearest Israeli city. The Israeli soldiers, exposed in that area, needed primarily to shield themselves, mostly in 1967–1970 and during the 1973 showdown. Yet Israel refused to withdraw its troops from the Suez Canal, even temporarily, because of political and military constraints. One of them was the concern that Egypt might obtain a foothold

**Israeli troops in a bunker near the Suez Canal during the war of attrition. (Photographer: Moshe Milner; Source: Israel's Government Press Office)**

on the east bank of the Suez Canal. This grip in Sinai could have served as a springboard for an Egyptian offensive, or it could have been the first stage in a political campaign that would have undermined Israel's control in the peninsula.

During the war of attrition in 1967–1970, Egyptian troops, in relatively small numbers, reached the east bank many times. Their missions were laying down ambushes, raiding Israeli posts, and so forth. They did not stay anywhere on the east bank, but they could have. The IDF would have quickly destroyed the isolated Egyptian force; yet Egypt could have tried this incursion again and again, figuring their presence on the east bank might attract international attention.

Either way, the IDF strove to push back or annihilate immediately any Egyptian force that crossed the Suez Canal. The IDF implemented forward defense, which was reasonable against limited assaults in a low-intensity war. Yet if Egypt had commenced a massive offensive, as it did in 1973, the IDF would have had to rely on defense in depth, enabling its troops to employ maneuver warfare, where they had an edge over the Egyptian military. But in the low-intensity war of 1967–1970, and in the 1973 high-intensity war as well, the IDF stuck to forward defense, which

**Israeli paratroopers in the west bank of the Suez Canal during the 1973 war. (Photographer: Ron Ilan; Source: Israel's Government Press Office)**

offered almost no flexibility. Israeli forces depended on one line of defense, which could have collapsed because of one weak link. Indeed, in 1973 the Egyptian military penetrated the Israeli line in many sectors, which ultimately brought down the entire line of defense.

In both the high- and the low-intensity wars until 1967, many of Egypt's soldiers were unaccustomed to fighting far away from their homes (about 200 kilometers) in the north of Sinai and in the Gaza Strip. Under those circumstances, at least some of them might not have felt compelled to do their best, and their morale might have declined. In the low-intensity war in 1967–1970 and in the high-intensity war of 1973, however, Egyptian troops fought on the two banks of the Suez Canal, much closer to their cities, towns and villages, which contributed to their motivation.

## Hybrid/Low-Intensity Wars, 1978–2005

On 10 March 1978, the IDF launched Operation Litani against the PLO in the south of Lebanon[56] after a deadly Palestinian assault in the Tel

Egyptian SA-2, an anti-aircraft missile, in the 1973 war. (Photographer: Yigal Tomarkin; Source: Israel's Government Press Office)

Aviv area.[57] Before the 1982 war the PLO believed that in case of an Israeli attack, it would repeat the pattern of the Litani operation.[58] Yet in 1982 the scope and the goals of the Israeli offensive were much bigger than they had been before. In the era of high-intensity wars, sometimes Egypt's military, as the PLO did in 1982, had assumed (wrongly) that the next Israeli offensive would be like the last one. For example, in 1967 Egypt's military thought that the IDF would concentrate its main effort in the same sector it attacked in the former high-intensity war between them, but this was not the case.[59]

The IDF worried that the need to deal with the Palestinian uprising, which started in 1987, would be at the expense of preparing for a high-intensity war against an Arab alliance.[60] In 1992, improvements in the IDF's ability to cope with the Palestinians, such as deploying special units, made it easier to reduce the number of troops in the West Bank and the Gaza Strip. Still, the uprising continued to be a burden on the IDF as far as getting ready to confront Arab militaries in a high-intensity war.[61] It was a familiar problem from former low-intensity wars; yet the length and scale of the Palestinian mutiny in 1987–1993 made it quite a challenge for the IDF.

Israeli soldier works on a security fence at the Golan Heights, on the border between Israel and Syria, 1974. (Photographer: Ya'acov Sa'ar; Source: Israel's Government Press Office)

At the end of July 1993, during a vast operation against the Hezbollah in Lebanon, the workload in the IDF's chopper squadrons sometimes resembled a war effort. The air crews gained combat experience, but the risk they faced was quite minimal.[62] Although the Hezbollah started then to demonstrate its hybrid capabilities, its fighters were not sufficient to jeopardize Israeli helicopters, despite the fact that they flew in a low altitude where they were exposed to even small arms fire. This kind of clash helped the IAF, in a limited way, to improve and test its air crews, as long as they did not get accustomed to the relatively minor threats facing them at that time. In a high-intensity war, the anti-aircraft defense would have been much more lethal.

In September 1996 most of the IDF's attention was focused on the Hezbollah in Lebanon, while the clash with the Palestinians (that is, the PA) during that month brought about a fundamental change.[63] The IDF understood that its men were not ready to deal with the burst of September 1996, because they were too busy with a hybrid/low-intensity war in Lebanon. The IDF therefore proceeded to invest more resources in preparing for another collision with the PA, one that might be much longer. This expected confrontation actually happened in 2000 and lasted five years.

On the first day of the 1967 showdown, when Israel's central front with Jordan was ignited, Israel concentrated its efforts against Egypt. Although Israel was aware that the Hashemite kingdom might confront it, the Israelis were surprised by the scale of the Jordanian operations.

In the clash with the PA in September 1996 the IDF was somewhat shocked, more so than in 1967. As in 1967, in 1996 Israel was involved in an ongoing fight on one front while its military had to deal with another front. In 1967 the IDF responded swiftly and successfully, according to its well-known flexibility, while in 1996 many in the IDF were rather disappointed by its performance. One reason for that was the threat to Israel. In 1967 it was quite worse, so the IDF was on full alert, including on the Jordanian front. But in 1996, following the Oslo Accords, the IDF did not consider the PA a rival. The collision in 1996 caused the IDF to change its perspective about the PA. Israel understood that, apart from an ongoing hybrid/low-intensity war in Lebanon, a low-intensity war might start on another front, perhaps against the PA (which was supposed to be relatively friendly). Therefore, since 1996, the IDF had to be ready to handle two fronts at the same time.

The PA is based on the PLO, Israel's sworn enemy from the mid-1960s until the early 1990s. In the 2000–2005 confrontation the PA (the PLO) returned—to a large extent—to be Israel's foe. In 2000–2005 the PLO was even more dangerous because it gained a foothold inside Israeli territory, one of its old goals. However, that also made the PLO more vulnerable. Its bases in Jordan until 1970 and in Lebanon until 1982 were within reach of the IDF, but it was not that easy for the latter to attack them since they were inside a sovereign Arab state. Jordan and especially Lebanon were not strong enough to hold back an Israeli offensive, but an Israeli operation— let alone a major one—within their territory was a complicated move. It might have caused a high-intensity war with other Arab states. Following the Oslo agreement, the PA had complete control in its "A" zones, and a penetration into them was, for Israel, a problematic undertaking. Yet the PA was not immune to attack, since it was not a state or part of an Arab state. Furthermore, PA areas were very close, and sometimes surrounded by the IDF, which was an obvious military advantage for the latter.

For the IDF, fighting the Palestinians in 2000–2004 was at the expense of training for a high-intensity war.[64] In late May 2004 four Israeli brigades, under one command, participated in an operation at Rafih in the Gaza Strip.[65] It was quite rare for so many brigades to be concentrated together in this low-intensity war against the Palestinians. The IDF allocated far fewer troops to most missions in that confrontation.

The offensive in Rafah highlighted the IDF's performance in controlling four brigades, an experience that would help it in a high-intensity war as well. Yet that maneuver in 2004 was not against a conventional military, since the Palestinians were lightly armed. It was an example of the many constraints of the 2000–2005 collision that made it harder for the IDF to prepare for a high-intensity war.

The IDF used night warfare in order to surprise its foe.[66] Night vision devices were a necessary tool in low-intensity warfare in 1967–1970, as in ambushes on the Jordanian front, as part of the struggle against Palestinian fighters who tried to infiltrate.[67] Yet Israeli troops were not able to fully exploit night warfare in both the 1973[68] and the 1982 high-intensity wars.[69] During the low-intensity war of 2000–2005, advanced measures of night vision gave Israeli troops an edge, such as in raids in urban areas. This experience could also have served them well in a hybrid and high-intensity war.

## Hybrid and Low-Intensity Wars Take Center Stage, 1982–2014

During recent decades, a conventional collision between Israel and Egypt would have been a typical high-intensity war.[70] Syria, like Egypt,

**A clash between an Israeli security force and Palestinians in the West Bank, 2000. (Photographer: Avi Ohayon; Source: Israel's Government Press Office)**

has a conventional military, but since the late 1990s Syria's military buildup has been based less on air force and armor, classic measures of a high-intensity war, and more on long-range surface-to-surface missiles and anti-tank and anti-aircraft missiles. In a way, Syria's military became a kind of huge hybrid force. Therefore, a large part of a clash between Israel and Syria would have been like a war between Israel and a much smaller hybrid force such as Hezbollah—that is, exchange of fire with various missiles, rockets and shells, and so on—and a collision between Israeli land units and anti-tank fire, mines and IEDs.

Until the 1980s Israel's main national security challenge was a high-intensity war, which caused the IDF to develop large armor formations— first a division and subsequently a corps. Since the 1980s the tanks have played a significant role in hybrid wars, and sometimes in low-intensity wars as well; yet the number of tanks involved has been much smaller. Most of the time the IDF did not need armor corps, divisions, brigades or even battalions—only companies or platoons.

The infantry in the IDF has become more valuable because of the growing importance of urban warfare in the West Bank, the Gaza Strip and Lebanon. The terrain in the latter is mountainous and rugged, and it also requires the infantry, as the maneuverability of tanks is more restricted there, and tanks are often stuck in narrow routes, exposed to enemy fire. The infantry's edge is therefore moving by foot, like guerrilla fighters, although the heavy weight the IDF's infantry soldier carries— weapons, equipment, ammunition, body armor, and so on—makes him much less flexible.

In the era of high-intensity wars in the Middle East (1948–1982), the IDF's infantry also faced a foe that possessed massive firepower, including tanks and heavy artillery, which guerrilla and terror organizations do not have. Still, hybrid foes like the Hezbollah and Hamas inflicted casualties with their sophisticated anti-tank missiles and IEDs, so IDF's infantry needed all the protection it could get, including well-armored vehicles like those that are based on a tank, the Namer and Achzarit.

The IAF adapted to hybrid warfare. Hezbollah and the Hamas did not possess an air force that needs to be destroyed, as Egypt did in 1967. Nor did those hybrid forces assimilate a lot of advanced anti-aircraft missiles, like those the IAF had to deal with starting in the late 1960s. The IAF thus had air superiority, but it could be jeopardized by low-altitude weapons and light arms (mostly by shoulder-fired anti-aircraft missiles).

The IAF bombed Hezbollah's headquarters in Beirut in the 2006 hybrid war,[71] which is, in a way, similar to the strike against Syria's general staff building in Damascus in the 1973 high-intensity war. The IAF also provided close air support. While in high-intensity wars the IAF had to be careful not to hit Israeli tanks located next to Arab tanks, in hybrid wars the IAF had to avoid targeting Israeli infantry when the latter was near enemy fighters. This was also a problem when using Israeli artillery, despite all its accuracy.

Since 1982 the IDF has taken various steps in order to be prepared for a high-intensity war, such as assimilating new planes, tanks, artillery, and so forth. Yet, during the last three decades, Israel's main security challenge has been mostly low-intensity wars. As a result, Israeli officers were too much focused on counter-insurgency. This was a problematic development. The 2006 war in Lebanon proved how difficult it was for the IDF to adjust to fighting a hybrid war after several years of clashing with the Palestinians in a low-intensity war in the West Bank and the Gaza Strip. If Israel's enemy in 2006 had been an Arab military like that of Syria or Egypt, the IDF's troubles might have been worse. Those Arab militaries were more exposed to Israeli firepower than a hybrid rival like the Hezbollah, but they could have struck Israel and its forces with much more lethal and destructive blows. For Israel, the ramifications of a high-intensity war might have been more negative than the consequences of hybrid and low-intensity wars, because of the scale of casualties and damages to civilian and military objectives.

The length and scope of the collisions with the Palestinians in 1987–1993 and 2000–2005 forced the IDF to allocate a large percentage of its soldiers to carry the burden of the fight. As a result, many of its troops had much less time to practice for a high-intensity war. It was a severe problem, not only because training for high-intensity war was often the only way to be fit for it but also because the drills for this kind of confrontation took a lot of time and resources due to the scale and complexity of the maneuvers. Preparing Israel's troops for low-intensity wars usually required shorter and simpler drills, or even just on-the-job training. Furthermore, low-intensity wars often lasted for years, so a unit unfamiliar with this type of conflict could be sent first to a quiet sector and then, in due course, to a more active one. By contrast, the pressure and scope of high-intensity war required that as many outfits as possible be qualified for this kind of confrontation, since they could be thrown immediately into the storm. Those constraints emphasized the need to invest in making the IDF ready for a high-intensity war.

The IDF gathered about 85,000 reserve troops during the war against

**Israeli troops on a street on the West Bank, 2002. (Photographer: Boaz Mesika; Source: Israel's Government Press Office)**

the Hezbollah in 2006 and around 36,000 reserve soldiers in April 2002 during a vast operation in the West Bank.[72] Therefore, ever since 1982, Israel's biggest mobilizations of reserve units participating in combat were in its hybrid war in Lebanon and at the peak of its low-intensity war against the Palestinians. The IDF continued to depend on its reserves, as in the era of high-intensity wars, but in much smaller numbers, since the foe was not as strong as conventional Arab militaries. This change allowed the IDF to give selected reserve units the best training, equipment and manpower (at the expense of other reserve units).

Fifty-six thousand reserve troops were gathered during the clash between Israel and the Hamas in November 2012.[73] They were, however, kept out of the campaign, which was conducted solely by air. Israel had never before had a high-intensity war in which the reserves did nothing on the ground while air power did all the work. Yet there were occasions during the era of high-intensity wars in which IDF's reserves were called in but did not join the skirmish. In the 1956 war, reserve forces on the Jordanian and Syrian fronts did not see any action, since the battles were only fought on the Egyptian front. In another event in February 1960, Egypt and Israel reinforced their deployments near their respective borders, but the reserves did not fight at all, since the crisis ended peacefully.

Following the Oslo agreement, the PA could have created a kind of army that might have invaded Israel and, by this means, disrupted the assembling of the reserves.[74] Another threat to Israel was a barrage of long-range missiles and rockets on its rear, where its population—that is, its reserves—are located. This kind of strike was expected from the 1980s, mostly at the beginning of a war with Syria, as part of a Syrian surprise attack in the Golan. The Hamas and especially the Hezbollah have acquired a similar option, but a more limited one. Like Syria in a high-intensity war, those non-state organizations might have delayed an Israeli offensive against them in a hybrid war.

Israel produced the "Iron Dome" system to shoot down rockets.[75] However, this kind of active defense might not have destroyed all the rockets, particularly those that were launched in huge waves against Israeli cities. Such an attack would be expected at the beginning of a war, when Israel's hybrid enemy was in full strength, as far as the amount of rockets it had. At that stage IDF's reserves would be at their homes, working places, and so on, unless the IDF had a warning that would prompt it to gather the troops in time.

Enemy rockets hitting their targets could cause confusion and even panic among some of the reserves. The same might have happened in the era of high-intensity wars if dozens of Arab bombers had broken through Israel's jets and air defense. In fact, Arab bombers might not have necessarily dropped more explosives in a high-intensity war than Arab rockets in a hybrid war. Either way, for the IAF (certainly before the campaigns of November 2012 and July–August 2014), intercepting rockets was a bigger challenge, because it represented uncharted waters, compared with bringing down Arab bombers. However, the rockets and missiles of the age of hybrid wars have exposed Israel to strikes no less than bombers in the era of high-intensity wars. The entire country has continued to be covered by enemy fire. Rockets and missiles could hit any spot, as Arab planes could have done in the past.

The manpower of the IDF has long been based on reserves and regulars. The reserves have had traditional advantages over the regulars, because they have served for several decades, many of them from their early 20s. Since every one of Israel's high-intensity wars occurred between 6 and 11 years after the last one, reserves in their 30s and 40s were thus more experienced than regular soldiers, who did not have the same background and seniority in military matters, since they served only three years. Since 1982, there have been no high-intensity wars, but reserves have continued to accumulate experience in hybrid and mostly low-intensity wars. This should have made it easier for the reserves to adjust to combat; yet they came from civilian life, while regular troops were already accustomed to performing as soldiers. Furthermore, over the years, when reserves were sent to perform tasks such as guarding bases, it was a waste of precious time since, on average, the reserves were on active duty for only a few weeks each year. It was better to use this short period to improve their skills in fighting a hybrid or high-intensity war by exercising, assimilating new weapon systems, studying combat doctrines, and so forth. Without such drills and studies, the IDF was deprived of a measure for estimating the effectiveness of the many reserve units that the IDF was based on.

The top brass, aware of the shortcomings of the reserves, but lacking a better option, might have assumed that the combat background of the reserves would be sufficient if there was a confrontation. This unpreparedness of the reserves was a risk in the case of a hybrid or a high-intensity war. In low-intensity wars it was usually less of a problem, since the demands of such warfare (and particularly the danger to Israel) were not that severe. For example, the reserves could have accomplished their missions in a low-intensity war without learning about the latest weapon system the IDF had acquired, like a new tank, if it was not used much in this kind of clash.

Regular troops, like the reserves, also perform routine day-to-day security assignments, but the former are in service all year long, so they are much more available for training. They are also young men, most between the ages of 18 and 21, while the reserves are older and often married with children. It was therefore no surprise in the border wars from the 1950s to the 1970s, particularly during raids, and also in the low-intensity war in Lebanon in the 1990s, that the regulars were much more dominant than the reserves. In the low-intensity wars with the Palestinians since 1987 and in the hybrid wars in 2006 and 2008–2009, there was more balance between the reserves and the regulars as far as their contribution went.

The average Israeli soldier could not have performed well without a firm belief in the IDF's goals, and without being aware of the importance of its tasks in the larger picture of Israel's national security policy. This was crucial in times of crisis and more so during combat. In the era of high-intensity wars, the familiar battlefields, mostly the Sinai, were quite desolate areas, and the IDF confronted conventional militaries like itself, not civilians (including armed ones). This fact spared the Israeli troops difficult ethical dilemmas, which could have been devastating to their morale, as manifested during the long and exhausting low-intensity wars that have occurred since the 1980s.

Besides the 1948–1949 war, Israel's high-intensity wars were usually confined to specific areas, mostly the Sinai and the Golan, that were far away from Israel's population centers. This was also the case during border wars like those in 1967–1970, as well as the hybrid/low-intensity war with the Hezbollah that went on from the early 1980s to 2000. This was one of the main reasons why the Israeli public was willing to tolerate this ongoing Sisyphean conflict. Although Israeli civilians living in the West Bank and the Gaza Strip have ever since the late 1980s known and felt they were in a battlefield during times of conflict, the majority of the Israeli people have lived normally, except for terror attacks (particularly when suicide bombers exploded in their cities). Such attacks did bring the war into the homes of many more Israelis, but it was not as bad as a massive Arab invasion in a high-intensity war, in which Arab tanks could have torn apart Israeli streets and houses.

Since the early 1980s Israeli society has been in the process of being worn down by the ongoing conflict and seemingly endless security problems. However, Israel and its military have not lost their determination to survive, nor have their rivals. The latter were often deterred, but on some occasions Israel's successes caused a fierce backlash. For example, the hybrid war of 1982 was a strategic blow to the PLO, but the confrontation gave way to the rise of the Hezbollah. From an operational point of view, at the beginning of the low-intensity war of 2000, the strong and effective response of the IDF infuriated the Palestinians and drove them to commit suicide attacks.

# 5

# Fire in the Backyard

## Israel's Campaigns in the Gaza Strip in 1956, 1967, 2008–2009 and 2014

After the 1948–1949 war, Israel and Arab states—such as Egypt—signed an armistice. As a result, the area that became known as the Gaza Strip was created and came under Egyptian control. However, it was far from the center of Egypt and, to a large extent, loomed as a military threat in Israel's backyard. Most of the Gaza Strip is surrounded by Israel, the skies over the Gaza Strip have been in range of Israeli planes, and the coast of the Gaza Strip on the Mediterranean Sea is quite close to Israeli naval bases.

Egypt (the ruler of the Gaza Strip from 1949 to 1967) and the Hamas (which ruled there since 2007) were both hostile toward Israel. In fact, the Gaza Strip has served as a base of operations against Israel. In the high-intensity wars of 1956 and 1967 the Gaza Strip could have been a springboard for a possible Egyptian offensive or Palestinian infiltrations. In the hybrid war of 2008–2009 and 2014, that area was used for artillery fire against Israeli targets, mostly civilian ones. As a result, the IDF launched a major attack against the Gaza Strip in 1956, 1967, wars of December 2008–January 2009 and July–August 2014. The IDF relied on its combat doctrine—attacking with ground and air forces—while taking into consideration factors such as the terrain of the battlefield and the strength of the enemy.

## The Ruler of the Gaza Strip

Egypt was an independent state in 1956 and 1967. Hamas in 2008–2009 and 2014 controlled the Gaza Strip the way a regular government does. Both Egypt in 1956 and 1967 and the Hamas in 2008–2009 and 2014 were not willing to recognize Israel's right to exist. Israel accepted Egypt as an independent state with a grip on the Gaza Strip, which was not annexed to Egypt. Unofficially, Israel also tolerated the Hamas dominion in the Gaza Strip, but in 1949–1967 it held Egypt accountable for any attack launched from the Gaza Strip. From 2007 onward Israel considered the Hamas responsible for any hostile act that came out of the Gaza Strip.

In the wars between Israel and Egypt in 1956 and 1967, the Gaza Strip was just a tiny part of Egypt, situated more than 200 kilometers from the heart of that country, and considered a remote outpost and part of the Egyptian defense of the Sinai. Egypt's problem was losing the Sinai to Israel; it could afford to lose the Gaza Strip (as indeed it did in those wars) without putting at immediate risk major cities such as Cairo. For Hamas, however, the challenge was much bigger, since in the confrontations of 2008–2009 and 2014 the Gaza Strip was all they had.

In 1953 the planning branch of the IDF estimated that the United States might tolerate Israel conquering the Sinai.[1] Yet after the 1956 war Israel actually faced U.S. sanctions, which forced the former to retreat from the peninsula, and the Gaza Strip as well, in March 1957.[2] After the 1967 showdown the United States rejected attempts to coerce Israel into handing over occupied territories, including the Gaza Strip, until a peace treaty was reached.[3] Israel indeed gave up about 80% of the Gaza Strip in 1994 as a result of the Oslo Accords with the PA. Later on, in 2005, the IDF completely withdrew from the rest of the Gaza Strip, including small parts of it that were later seized in the confrontations of 2008–2009 and 2014. This was done without any official arrangement, let alone a peace agreement with the other side.

In the 1956 war, Israel strove to bring down Nasser, but he stayed in power.[4] In 1967 Israel hoped again for the same outcome. Following his defeat, Nasser indeed resigned, but then he backtracked and remained in office.[5] In the days before the confrontation of 2008–2009, Israel's deputy prime minister, Haim Ramon, called for the end of the Hamas.[6] In the war of 2014 Israel's strategic affairs minister, Yuval Steinitz, strove to topple the Hamas by conquering the Gaza Strip for a short period.[7] This move was relatively easy to achieve, in comparison with overthrowing Nasser in 1956 and 1967. Israel, however, avoided such a move, fearing that a chaotic alternative to Hamas might be even worse.

Egypt in the 1956 and the 1967 wars was left alone, since its allies, Syria and Jordan, did almost nothing to ease the situation for Egypt when Israel invaded Sinai and the Gaza Strip. In the confrontations of 2008–

2009 and 2014, the Hamas did not receive any real Arab support either. Some Arab states, such as Jordan, even considered Hamas a threat, the way they saw Nasser in 1967. Hamas in 2008–2009 and in 2014 was much less powerful than Nasser in 1967, but, from a Jordanian perspective, both Nasser and Hamas were dangerous to the stability of the Hashemite kingdom.

Before the 1956 war, the United Nations served as a broker between Israel and Egypt during their border clashes. The international organization sometimes helped to reduce tension and obtain a ceasefire.[8] In June 2008 Egypt managed to mediate between Israel and Hamas following the friction between them.[9] But, as with the United Nations in the 1950s, Egypt only succeeded in postponing a major collision, not preventing it.

## *The Gaza Strip as a Base Against Israel*

In late 1955 Egypt blocked the Tiran Straits in south Sinai to Israeli ships. The IDF had a plan to seize that area. In response, the Egyptian military might have attacked from the Gaza Strip,[10] which, due to its location, is like an enormous bulge inside Israel, thus providing an ideal launching point for an assault. In the early and mid-1950s, as part of their low-intensity war effort, Palestinians penetrated into Israel from the Gaza Strip. The rate of this infiltration slowed down before the 1956 war, but it could have resumed full-scale in a short time. Capturing the Gaza Strip would have prevented that. Furthermore, Israel had to conquer that area anyway, since the Gaza Strip was a threat to the flanks of Israeli troops when they advanced inside the Sinai during their high-intensity war against the Egyptian military. Seizing the Gaza Strip was therefore required due to constraints of both high- and low-intensity wars.

Before the 1967 war, Israeli military intelligence estimated that Arab states might start a wave of terror and guerrilla raids from the Gaza Strip and other locations.[11] However, as in 1956, seizing the Gaza Strip was above all part of the campaign in Sinai against Egypt's military.

In later years, the Hamas continued to penetrate into Israeli territory from the Gaza Strip, and on June 25, 2006, two Israeli soldiers were killed and one, Gilad Shalit, was captured.[12] Furthermore, about 6,000 rockets and mortars were fired from the Gaza Strip into Israel during the years 2005–2008.[13] In the campaign of 2008–2009 Israel's primary goal was not so much to cut off infiltrations from the Gaza Strip, but rather to stop the Hamas from shelling the southern part of Israel. In the war of 2014

Israel strove to stop the firing and destroy tunnels that led into its territory.

Although Israel was forced to retreat from the Gaza Strip a few months after the 1956 war, the border with the Gaza Strip was quiet at least until 1962. This proved that the major offensive in 1956 had achieved the desired results. Following the hybrid campaigns of 2008–2009 and 2014, Israel left the areas it had seized in the Gaza Strip. Those two Israeli operations did not deter the Hamas from rearming, but, particularly after the clash in 2014, there were almost no assaults or shelling directed at Israel. It showed the effect of a large offensive in a hybrid war, which resembles the outcome of the attack in the high-intensity war of 1956.

## The Importance of the Terrain

The Gaza Strip is a tough area to defend, since its width is at most 13 kilometers, and there is a need to protect all of it, since it has no key sites from which to control the entire area.[14] Egyptian forces that were deployed there in 1956 and 1967 had no depth. In the wars of 2008–2009 and 2014, only a few kilometers separated the strongholds of the Hamas in the city of Gaza and the border with Israel. In the war of 2008–2009 the IDF split Hamas' forces by cutting the Gaza Strip in half in a matter of hours.

In 1956, 1967, 2008–2009 and 2014, the defender of the Gaza Strip did not possess fortifications strong enough to halt the Israeli offensive. Furthermore, the open desert terrain of the Gaza Strip made it easier to invade from different directions, which forced the defender to spread its resources to several sectors. The coast was another possible entry point, but the IDF never tried to land a major force on the shores of the Gaza Strip. The advantages the IDF had in the other sectors of that area were probably sufficient as far as routes for invasion. The IDF also wished to avoid an amphibious assault in which its officers had no combat experience.

At the beginning of the wars of 1956, 1967, 2008–2009 and 2014, the IDF had almost completely surrounded the Gaza Strip by land and sea, leaving only the Rafiah sector as a gate to Sinai. For that reason, the IDF invaded the Rafiah sector, cutting off the Gaza Strip from Sinai in 1956 and 1967. In the wars of 2008–2009 and 2014 the IDF did not penetrate the Rafiah sector, but Israeli planes bombed Palestinian targets—such as tunnels running below the border—through which the Hamas received military equipment from Sinai.

In the Gaza Strip there were 200,000 Palestinians in 1956,[15] 350,000–400,000 in 1967,[16] 1.5 million in 2008[17] and 1.8 million in 2014. These growing numbers have emphasized the difficulty of avoiding civilian casualties and the increasing importance of urban warfare, mostly demonstrated in the confrontations of 2008–2009 and 2014.

## Israeli Military Doctrine

The Israeli military doctrine is based on reaching a fast decision. The entire Gaza Strip was seized in one day in 1956, and in three days in 1967. In 2008–2009, however, even after 22 days of combat, and in 2014, even after roughly two weeks of battles, only part of the Gaza Strip was taken. This sharp contrast was due to the fact that in 1956 and 1967, Israel had concluded it would occupy the Gaza Strip, while in the entire confrontations of 2008–2009 and 2014 no such decision was made.

The IDF assumed in 1956 that there might be a war, so in the months before the confrontation almost all its reserve troops were drilled.[18] In the war itself they were mobilized in a few days.[19] The crisis leading to the 1967 war came as a surprise; yet the IDF had three weeks to get ready, including its reserves.[20] In 2008–2009 the IDF expected a clash and had trained properly in advance.[21] The IDF also had several days in which to call up its reserves when the confrontation started (as was also the case in 2014). In all those collisions Israel had time to train its troops and to deploy them.

Ever since it was established, the IDF has had to prepare for war on several fronts, mostly the Egyptian, Syrian and Jordanian ones. In the 1956 war the IDF was able to allocate 10 of its 16 brigades, most of them reserve units, to the Egyptian front that included the Gaza Strip. This was made possible by the fact that Egypt's allies, Jordan and Syria, stayed out of that war. In 1967 Syria and Jordan did very little to assist Egypt when it was attacked by Israel, which allowed the IDF to penetrate into Sinai and the Gaza Strip with three of its four divisions, most of them reserve forces. In 2008–2009 and in 2014, Israel's foe, the Hamas, not only stood alone but also was much weaker than the Egyptian military had been in 1956 and 1967. The IDF concentrated its efforts on the Gaza Strip without even needing to call in most of its reserves.

Prior to the 1956 and 1967 wars, Israel created fortifications in the Negev, in case of an Egyptian invasion from Sinai or from the Gaza Strip. In 2008–2009 and 2014, Israel had a fence, observation posts, and so on along the border with the Gaza Strip. However, as in 1956 and 1967, strictly defensive measures were not a favorable solution, mostly because of the high cost of constructing fortifications for troops, and shelters for civilians, in the south of Israel. Furthermore, such an approach allowed Israel's foe to have the initiative. As a result, in 1956, 1967, 2008–2009 and 2014, Israel eventually concluded that it had to go on the offensive.

In 1956, Egypt assumed that Israel might exploit the British-French attack on Egypt in order to confront it. Yet Egypt did not foresee the scale of the Israeli attack, since it was not aware of the secret treaty between Israel, France and Britain. In 1967 as well, in spite of high tension before the war, Egypt was surprised by the Israeli offensive. In 2008 the Hamas anticipated that Israel would respond to the shelling of its civilians, but it was still quite astounded by the magnitude of the Israeli action. In 2014, although the Hamas was prepared for combat, it still failed to overcome the Israeli defense and stop its offensive movement. Therefore, in all four confrontations the defender of the Gaza Strip anticipated a clash but failed to predict its scope and/or the full capabilities of the IDF.

The 1956 war started with the landing of an Israeli battalion parachuting deep inside Sinai, as part of the Israel/France/Britain alliance plan. The aim was to give the two European powers an excuse to invade Egypt. The 1967 showdown began with a massive Israeli air raid on Egyptian airfields. The first stage of the 2008–2009 confrontation was a series of devastating air attacks on the Hamas. The first step in the wars of 1956, 1967 and 2008–2009 was therefore a move by Israel that shocked its foe. In 1967 and 2008–2009 Israel launched a heavy air bombardment that had an impact on the rest of the war. The vertical flanking from the air in 1956 was more of a political maneuver than a military one; yet it, too, had a major effect on the entire campaign, for it helped Britain and France attack Egypt. Unlike those confrontations, the IDF did not initiate any first strike in the clash of 2014.

Infiltrations into Israel from the Gaza Strip in the first half of the 1950s persuaded Israeli soldiers that conquering the Gaza Strip in 1956 was justified. In the 1967 showdown the common belief in Israel was that the state was facing a possible attack from Egypt, which at that time controlled the Gaza Strip. In 2008–2009 and 2014 the ongoing artillery fire from the Gaza Strip (and, in 2014, the need to destroy the menacing tunnels) convinced Israeli troops that there was a legitimate reason for war, thus enhancing their motivation and will to demonstrate courage in combat. In all four wars, the high morale of the troops gave the Israeli government the backing it required to order an attack.

## The Israeli Ground and Air Forces

In 1956 and 1967 there was not much of a threat to the IAF in the Gaza Strip from Egyptian planes or anti-aircraft batteries. Likewise, in the confrontations of 2008–2009 and 2014 the IAF was not really challenged by the Hamas in the skies of the Gaza Strip, since Hamas had only shoulder-fired anti-aircraft missiles and heavy machine guns. This basically meant that the IAF had freedom of movement.

In the 1956 war, during the ground offensive in the Gaza Strip, the IAF launched very few air strikes. There was not much artillery support, either.[22] In the 1967 showdown Israeli troops penetrated Gaza city after five hours of shelling.[23] In the confrontations of 2008–2009 and 2014 Israel's land units received massive air and artillery support, throughout the entire campaign.

In the 1956 war Israeli tanks participated in conquering the Gaza Strip while dealing with a few anti-tank ambushes.[24] Moshe Bar-Kochva, who was a very experienced armor officer, claimed that in 1956 tanks proved to be the main force in conquering the Gaza Strip.[25] In 1967, in the area of Khan Yunis (a city in the south of the Gaza Strip), various anti-tank fire, mines, and natural and artificial barriers delayed and caused casualties to the 7th Armor Brigade.[26] In 2008–2009 the 401st and the 188th Armor Brigades penetrated successfully into the Gaza Strip.[27] In 2014 Israeli tanks were heavily involved in the ground offensive. In all four confrontations, Israeli armor played an important role in the battles in the Gaza Strip but not necessarily a decisive one. It contributed to the battles thanks to its speed, protection and firepower.

During the 1956 war in the Gaza Strip there was a lack of coordination between Israel's infantry and armor.[28] This was still a problem in the 1967 war.[29] In the clash of 2008–2009, the coordination between Israel's ground and air forces went off quite well,[30] and in 2014 it was even better. It was always a challenge to conduct joint operations between the main corps: armor, infantry, air force, and so forth.

## Military Deployment and Buildup in the Gaza Strip

The enormous arms deal Egypt made in late 1955 with the Soviet Union included hundreds of armored vehicles and jets. However, the

Egyptian military was not able to assimilate many of those weapon systems until late October 1956, when the war started. Israel had also acquired hundreds of tanks and jets during 1956; yet its leadership assumed this was not enough to deter Egypt.

Israel was worried that Egyptian armor could use the Gaza Strip as a jumping-off point to reach the heart of Israel, which was about 80 kilometers north. Although in late 1956 Egypt's attention was focused on the Suez Canal and not on Sinai, let alone the Gaza Strip, this situation could have changed quite fast. Therefore, Israel launched an offensive in 1956 and defeated the Egyptian military. In 1967 Israel was highly concerned, since the Egyptian deployment near its border was much bigger in comparison to 1956. The Egyptian military also had years to train with its weapon systems, and many of them were better than the ones it possessed in 1956. Egypt could have exploited all those advantages to commence an offensive from Sinai and/or the Gaza Strip, so Israel attacked and annihilated the Egyptian forces. Before the confrontations of 2008–2009 and 2014, a limited military response and diplomatic efforts failed to stop the fire from the Gaza Strip, and the Hamas continued to build its arsenal. This pushed Israel to start a major offensive.

In the 1956 and 1967 wars, Egypt armed Palestinians in the Gaza Strip.[31] Prior to the campaign of 2008–2009, Egypt did not deliver weapons and ammunition directly to the Palestinians in the Gaza Strip but, to a large extent, ignored the smuggling of weapons and other military supply from Egyptian territory, the Sinai, into the Gaza Strip. This process made the Hamas stronger and increased its confidence in defying Israel. The same thing happened in the years before the 2014 confrontation, although in the months leading up to that campaign Egypt did make a substantial (and successful) effort to block the smuggling from Sinai to the Gaza Strip.

In the 1956 war Egypt had the 8th Palestinian Division in the Gaza Strip. The troops were Palestinians, while their officers were Egyptians. The division, 10,000 men strong, had brigades, battalions and companies, but they were not operational formations. Those troops did not have many support weapons, and they were split into dozens of posts that were unable to assist each other. Their garrisons demonstrated some resistance but not enough to hold back the Israeli offensive.[32] In the 1967 war the 20th Palestinian Division in the Gaza Strip[33] put up a good fight in several places, but eventually its troops were beaten.[34] In the confrontation of 2008–2009 the Hamas had about 15,000 fighters, none of whom were a match for the Israeli military in terms of their skills, training, and so forth.[35] In the war of 2014, this scenario repeated itself.

During the 1956 and 1967 wars, the importance of the Gaza Strip as a springboard for Egypt was mitigated by this territory being vulnerable to Israeli attacks because of its location, length and width. Egypt did not wish to expose too many of its troops there, particularly since there were other priority areas (in 1956 it was the Suez Canal, and in 1967 the Sinai). The Hamas in 2008–2009 and 2014 did not have any alternative territory, so they had to make do with the disadvantages of the Gaza Strip.

# 6

# Fighting Hybrid Adversaries
## *Israel Versus the PLO in 1982*
## *and the Hezbollah in 2006*

The 2006 confrontation between the IDF and the Hezbollah was a hybrid war.[1] Hezbollah's action represented "one example of a future hybrid threat that encompasses the essence of hybrid warfare."[2] In 1982 Israel collided in Lebanon with another hybrid rival—the PLO. Those hybrid collisions included strategic aspects such as setting realistic goals and operational factors, some of which were urban warfare, firepower versus maneuver, the importance of training, and so forth.

## *The Israeli Perspective: Strategic Aspects*

In the early 1980s Ariel Sharon, Israel's powerful minister of defense, considered the PLO a danger to Israel.[3] The border between Israel and Lebanon (more specifically, the PLO) was relatively calm in the months before the 1982 war. Nevertheless, Sharon and others in Israel assumed it was only a matter of time before the PLO renewed its terror and guerrilla activities, possibly with greater vigor. Israel therefore wished to defeat the PLO, which led to the 1982 war. Similarly, prior to the 2006 war, the border between Israel and Lebanon—that is, the Hezbollah—had been quiet for several years, but the Israeli leadership wished to disable the Hezbollah and prevent it from resuming its attacks in the future.

On 3 June 1982 the Israeli ambassador in London barely survived an assassination attempt. Three days later IDF started the war against the PLO in Lebanon. On 12 July 2006, the IDF began a war against the Hezbollah following a skirmish very near to the border in which two Israeli soldiers had been kidnapped. Therefore, in both 1982 and 2006 the spark

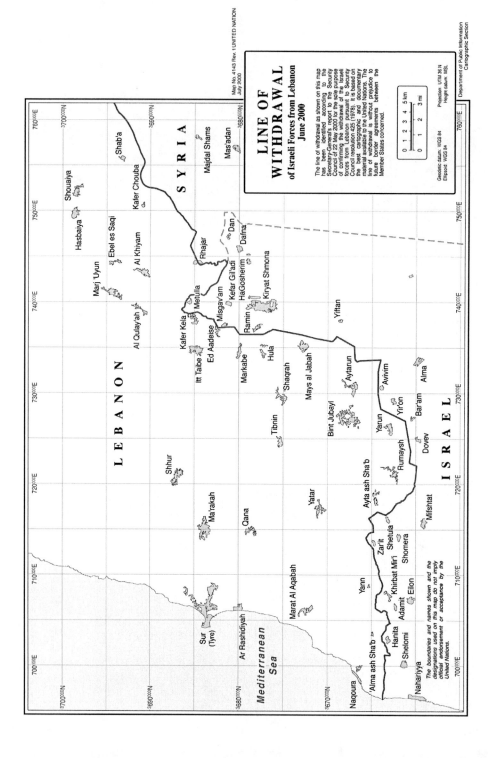

was a provocation against Israel. Those in charge of both the PLO in 1982 and the Hezbollah in 2006 did not seek to initiate a war. In 1982 the assault on the Israeli ambassador was not even planned by Yasser Arafat, the leader of the PLO. Yet it was an opportunity for Israel, on both occasions, to bash its sworn enemy in Lebanon.

In 1982 the Israeli war plan consisted of two general versions: a limited one with which the war started, and a much more ambitious one that was activated during the battles.[4] In the 2006 war Israel aimed high from the start of the war, but, as in 1982, it failed to achieve all of its objectives.

One of Israel's goals was the same in both wars: to remove completely the threat of missiles and rockets fired from Lebanon into Israel. In that sense there were at least nine years of quiet after those confrontations, but this challenge remains, and once again the problem stems from the Hezbollah.

Before the 1982 war there were talks between the Reagan administration and Israel about the situation in Lebanon. Israel could have assumed that the United States would allow the IDF to carry out a limited operation in Lebanon against the PLO.[5] Yet the extent of the Israeli offensive surprised the U.S. leadership, who did not wish Israel (and surely not their own country) entangled in a severe and unnecessary crisis.

In 2006, the Bush administration may have been notified in advance about the Israeli plans to confront the Hezbollah. In any case, the United States supported the Israeli offensive to bash the pro-Iranian Hezbollah.[6] The Americans assumed, as they had before 1982, that Israel would achieve a quick victory against a non-state organization that was also an enemy of the United States.

## *The Israeli Perspective: Operational Aspects*

In 1982 and 2006 there was a "strategy-policy mismatch" by Israel[7] and some tension between the Israeli government and military.[8] In both wars there were also some in the IDF who did not agree with the strategy of the war, if only in terms of its military aspects. Their reasons, among others, claimed lack of clarity about the objectives and often changed orders—especially in 2006—which confused and frustrated Israeli officers.

A year before the 1982 war, the Israeli government estimated that there might be a collision in Lebanon. In the year that followed, the IDF prepared intensively for such a campaign. Conversely, when the Israeli government decided to go to war on 12 July 2006, the IDF was surprised. Indeed, the war in 2006 revealed the lack of equipment, training and overall readiness of Israeli troops.[9] For example, "Undoubtedly, the actions of division 162, at Wadi al-Saluki, underscore the dismal state of the IDF's ground forces, particularly in conducting conventional maneuver operations."[10] In 1982, therefore, the IDF was better prepared to fight than in 2006, which explains why the IDF hesitated to launch a ground attack in 2006 and preferred to rely on firepower.

In the years before 2006, the IDF focused on confronting the Palestinians in the Gaza Strip and the West Bank.[11] The war in Lebanon in 2006 demonstrated how Israeli troops, such as reserve troops from the armor corps, were not well trained, since they had done mostly infantry missions in Palestinian areas during their regular service.[12]

In the years running up to the 1982 war, the IDF had to fill day-to-day security assignments as well. Those included handling matters in Palestinian territory then under Israeli control—namely, the West Bank and the Gaza Strip. Those tasks were also performed before 2006, but it seems they took their toll more at that time than prior to 1982, which enabled the IDF in 1982 to allocate more time and resources to train for the war in Lebanon.

In all wars fought since the early 1950s, the IDF has sought to gain a rapid victory—that is, to win in a few weeks and even days, as in the wars of 1956 and 1967. In 1982 the war between Israel and the PLO started on 6 June. It took the IDF about a week to get to the gates of Beirut, where the PLO held its ground until it withdrew on 21 August 1982. This outcome was an Israeli achievement, but it took more than two months to attain it. Furthermore, it was not much of a triumph, since Israel did not accomplish other goals it had aimed for. The IDF continued to clash in Lebanon with other enemies, mostly the Hezbollah, until 2000. The war in 2006 lasted 34 days, and at its end the Hezbollah was not clearly defeated, since the organization survived and has since kept its position in Lebanon.

IDF's combat doctrine called to transfer the battles as soon as possible into the opponent's land. Two of Israel's main operational goals were to annihilate enemy forces and to conquer vital territories. In 1982 Israel stuck to this approach. The IDF stormed into Lebanon and occupied large parts of it while destroying PLO units. In 2006, on the other hand, Israel captured much less ground and the process took longer than it had in 1982. Israel also bashed the Hezbollah; yet the IDF did not focus on this aim alone. The Israeli intention was to create the illusion that the foe lost without actually beating the enemy forces, and particularly without seizing their land. This reluctance of Israel to capture huge parts of Lebanon in

2006 was also due to its bitter memory of the 1982 war, when Israeli troops were bogged down in Lebanon for 18 years.

In 1982 the IDF preferred, during its land offensive, to rely on firepower and not on maneuvering.[13] In 2006 the IDF took this approach even further and tried to avoid a ground offensive altogether.[14] The reasons in both wars were the same: reducing as much as possible the number of Israeli casualties. In 1982 Israeli firepower helped save the lives of Israeli troops, when in about a week the PLO was pushed deep into Lebanon. This move secured the Israeli population from shelling and infiltrations. In 2006 Israel's firepower was not able to beat the Hezbollah, and for more than a month Israeli citizens were exposed to enemy fire. It was a grim result of the IDF's attempt to imitate a recent war run by Western states (mostly the United States): the 1999 campaign in Kosovo, which relied on airpower. Regrettably, the IDF chose to ignore the lessons of the invasion of Iraq in 2003, where a ground offensive was also essential.

In 2006 the IDF had more precision-guided munitions than in 1982, and they were supposed to limit collateral damage. Still, it was not easy, since the IDF had to pinpoint precisely the location of enemy positions, such as rocket launchers, often placed inside or very close to towns and population. No wonder there were incidents in which dozens of Lebanese civilians were killed by mistake.

About a year before the 1982 war, in mid-July 1981, there was a campaign that lasted 12 days. Israel bombed targets inside Lebanon while the PLO fired shells and rockets into the north of Israel, causing many of its civilians to flee to the rest of the country. Both sides endured casualties. There was a heated debate among the Israelis during and after this collision over whether to initiate a full-scale offensive in Lebanon.[15]

In April 1996 "Grapes of Wrath" was a major Israeli operation against the Hezbollah,[16] and the last big collision between the two sides before 2006, although skirmishes in Lebanon went on until 2000.

In 1982 the IDF deployed six divisions in Lebanon,[17] whereas in 2006 the IDF had five divisions deployed against the Hezbollah.[18] Looking back, this makes sense, for in 1982 a large part of the Israeli force penetrating Lebanon was needed to confront the Syrian military in case of a collision. In fact, Syrian troops stationed in Lebanon did encounter the Israelis on that occasion. In 2006, however, Syria was out of Lebanon, and there was only a small probability of a Syrian intervention, which—as it turned out—did not happen. Likewise, in 1982 and 2006 the divisions of IDF's northern command were on alert about Syria clashing with them in the Golan. But Syria avoided that in both wars.

In 1982 one Israeli division was sent to the city of Tyre and three others to the city of Sidon,[19] two strongholds of the PLO. Indeed, several divisions were allocated from the beginning of the war against the PLO, in spite of the Syrian presence on the same front.

Generally speaking, the IDF had an overwhelming force in Lebanon, which might have been a bit too much considering the strength of the foe, particularly the PLO, and the rugged terrain of Lebanon, which confined vehicles to narrow roads. It seems that in 1982 Israel wished to avoid the shortage of troops it had suffered during the former war in 1973; perhaps it even overcompensated for it.

In the 2006 war the IDF was unprepared to deal with the nature of the Lebanese terrain.[20] Before the 1982 war it was also vital to study the terrain.[21] It was a necessary lesson ever since the first Israeli campaign in Lebanon during the 1948–1949 showdown. Although the IDF had launched a limited attack in that war, it was clear that an advance by foot or with vehicles would be slowed down due to the hills and mountains of Lebanon.

In 1982, Israeli armor units "were almost irrelevant against the guerrillas of the PLO."[22] In 2006 anti-tank missiles in the hands of well-qualified crews proved to be effective against the IDF.[23] In the terrain of Lebanon the IDF had difficulties in maneuvering, which exposed even its most protected tanks to enemy fire. It was especially a problem in 2006, since Israeli armor troops were not properly trained.

In 1982 the IDF gradually took over Palestinian refugee camps in south Lebanon and split them into separate parts. In some of these places the IDF had a hard battle. Conquering the camps in the area of Tyre cost the lives of 21 Israeli soldiers.[24] In 2006 the IDF confronted the Hezbollah in places such as Bint Kbeil, a large town in south Lebanon. In one of the clashes, on 26 July, Israel's 1st Brigade was thrown into a tough fight and lost eight men.[25] Urban warfare proved, therefore, to be quite a challenge for the IDF in 1982 and 2006.

The IDF avoided vertical flanking in the 1982 war, fearing the cost of such missions.[26] At the end of the 2006 war, the IDF conducted vertical flanking; yet it was mostly symbolic.[27] The IAF had air superiority in both 1982 and 2006, an essential condition for airborne assaults. The rugged Lebanese terrain was another reason for sending troops by air to key spots, but the PLO in 1982 and the Hezbollah in 2006 had anti-aircraft weapons. Israeli helicopters were particularly vulnerable in low altitudes, which deterred the IDF from large-scale landing from the air in both wars.

In the 1982 war the 35th Paratroop Brigade carried out an amphibious operation on the beach of Awaly, near Sidon. The aim was to cut off PLO units in the south of Lebanon and encircle them.[28] As it turned out, this opportunity was missed, although the IDF faced relatively light resistance.[29] In 2006 the IDF had very few landing craft, preventing it from exploiting Lebanon's long shoreline. Furthermore, Hezbollah's coast-to-sea missiles, which almost sank Israel's flagship, could have disturbed landing as well.

## Why Did the PLO and the Hezbollah Stand Alone?

The PLO in 1982 and the Hezbollah in 2006 were non-state organizations that managed to establish themselves as independent players in the Arab world, although they needed support from Arab states. During the 1982 and 2006 wars Arab militaries could have intervened, forcing the IDF to split its efforts onto several fronts instead of focusing on Lebanon. However, Arab states abandoned the PLO in 1982 and the Hezbollah in 2006. Furthermore, in 2006 many Arab countries ruled by Sunnis actually looked forward to the collapse of the pro-Iranian Shiite Hezbollah.

In 1982 an Israeli move against the PLO in Lebanon might have been the first stage in a plan to weaken Syria's position in that country.[30] From Syria's perspective, Lebanon has always been not only part of Syria but also a shield, protecting key Syrian cities, including Damascus.[31] Therefore the Israeli attacks in 1982 and 2006 worried Syria. In 1982 Israel had seemingly focused only on the PLO but quite quickly discovered that it had to deal with not only a hybrid war with the PLO but also a high-intensity war against Syria. The confrontation between the two states was a limited one, and it did not relieve the Israeli pressure on the PLO. In 2006 Syria was officially out of Lebanon; yet Assad had close ties with the Hezbollah, a relationship that was stronger than the one Syria had with the PLO in 1982. An Israeli threat on the Hezbollah was, for the Assad regime, a major hazard. In 2006 a total defeat of Syria's main ally in Lebanon would have given its many foes there an opportunity to seize control, which would have further reduced Syria's influence in that country.

An anti-Syrian government might have gained enough confidence to collaborate with Western powers such as France and the United States against Syria itself, publicly presented as part of the global fight against terror. In spite of those possible ramifications, Syria did not rush to assist the Hezbollah and at most intervened indirectly in the campaign. Syria did not want to risk any collision with Israel. Furthermore, while in 1982 Syria enjoyed the backing of a superpower—the Soviet Union—in 2006 the Assad regime could have relied only on Russia (now a secondary power) and Iran (a regional power).

## The PLO and the Hezbollah in Combat

In 1982, although the PLO in south Lebanon had many posts fortified with bunkers, mines and fences, it could only hope to delay the Israeli offensive, not to stop it. Indeed, after 48 hours of combat, its men were ordered to withdraw to Beirut or the Lebanon valley. Ultimately its defense plans were not implemented. The Palestinians did not blow up bridges or mine the roads properly,[32] although stiff Palestinian resistance in places like Tyre and Sydon prolonged the fighting.[33] In south Lebanon refugee camps with dozens of bunkers were organized for battle independently. In one of them, Ein Eilwa, the battle lasted about a week,[34] longer than many collisions between the IDF and conventional Arab militaries in high-intensity wars. In Beirut the siege lasted more than two months, but not because the PLO held back the IDF. The delay was due to political reasons and Israel's fear of heavy losses among both its troops and the Arab population during a conquest of the Lebanese capital.

In 2006 the Hezbollah had fortifications and bunkers. The IDF assumed that the Hezbollah had relied on forward defense near the border, but in fact it was defense in depth.[35] Yet the Hezbollah did not possess the same level of firepower, fortification and manpower that the Syrian military had had on its front with Israel. In 2006 the Hezbollah slowed down the advance of the IDF, but, considering the full potential of the latter, it was only a matter of time before the defense lines broke. Israel's mistakes and hesitations in launching a major ground offensive don't mean Hezbollah's defense should receive more credit than it actually deserves.

In the 1982 war, psychological warfare was called into play in refugee camps, in order to convince the Palestinians to turn themselves in by promising them that they would not be harmed.[36] The IDF also struggled to tell the Palestinian warriors (including children with rocket-propelled grenades) from the noncombatants.[37] In areas such as Tyre and Sidon the population was exposed to danger because it was mixed with the PLO.[38] The IDF tried to distinguish between PLO fighters and civilians by encouraging the latter to leave their homes temporarily, so only combatants were left there. The IDF had a similar problem in 2006 when Hezbollah launched its rockets and missiles from towns and villages.

In the 1982 war the PLO was about a division strong, and in the process of transforming from a guerrilla group into a military formation that included battalions, dozens of tanks, and so forth.[39] Rafael Eitan, the chief of staff of the IDF in the 1982 war, claimed that this change worked in favor of the IDF, which prefers to deal with a conventional force and not with a terror organization.[40] During the war "PLO brigades were quickly scattered," which actually helped their men confront the IDF.[41] Most of the Palestinian troops of the "Castel" brigade, which had six battalions, tended to withdraw, and the entire unit disintegrated under the Israeli pressure.[42]

In 2006 the Hezbollah did not fall apart in face of the Israeli advance. Its command and control system functioned well, and its "territorial units were decentralized and operated autonomously" to begin with.[43] Similar to the PLO in 1982, Hezbollah fought in small details.

Until 2006, "Hezbollah was able to efficiently adjust its tactics and operational design. Its planning was simple and inspired."[44] Hezbollah fighters proved to be professionals,[45] and they demonstrated a high level of initiative and flexibility.[46] In 1982 many PLO fighters had also proved their courage and determination,[47] although the Hezbollah in 2006 was considered a tougher adversary than the PLO of the early 1980s.[48]

## Lebanon As a Base

In Lebanon, the PLO and the Hezbollah exploited the advantages offered by the fragile government of a torn country and a friendly local population that supported them. The PLO had the Palestinians, and the Hezbollah had the Shiites. The PLO in the 1970s, like the Hezbollah since the 1990s, gradually consolidated a mighty grip on Lebanon, which was a springboard from which to attack Israel with artillery fire and/or infiltrations.

The PLO had in south Lebanon a "quasi state."[49] In the mid-1970s Israel was somehow willing to tolerate PLO deployment south of the Litani River in Lebanon, as long as there were no units from Arab militaries.[50] (The latter represented much more of a threat to Israel.) In the same line of thinking, up until the war in 2006, Israel accepted the presence of another hybrid force, the Hezbollah, south of the Litani River, all the way to the Israeli border.

One of the Israeli goals in 1982 was to put an end to the military resistance of the Palestinians by crushing the PLO.[51] The PLO suffered a substantial setback and lost much of its base in Lebanon, but the organization survived.[52] In 2006 there was no clear winner.[53] The Hezbollah absorbed blows and heavy losses, but it remained on its feet and was not kicked out of Lebanon, not even from Beirut (although its presence there was severely damaged due to Israeli bombardments). This outcome was a bitter disappointment for Israel, since the Hezbollah managed to restore (and even upgrade) its capabilities in Lebanon.

Another Israeli aim in the 1982 war was to create a friendly regime in Lebanon[54] following its secret negotiations with its allies, the Lebanese Christians. This attempt to establish a pro-Israeli leadership, which would have helped Israel neutralize the PLO in Lebanon, failed. In 2006 Israel did not wish to bring about regime change in Lebanon, but it hoped the local leadership would restrain the Hezbollah. Such an endeavor did not succeed, due to the weakness of the Lebanese government.

For the Hezbollah, the conflict with Israel, as in the war of 2006, was both political and ideological. For the Palestinians, it was much more than that, mostly because hundreds of thousands of them were forced to leave their land in the 1948–1949 war between Israel and the Arabs. Hezbollah in 2006 and the PLO in 1982 were non-state organizations, but, like most Arab states, they refused to recognize Israel's right to exist. Still, the PLO's main goal was to use Lebanon to create a Palestinian state instead of Israel. Hezbollah's top priority, in spite of all its bombastic declarations of marching to Jerusalem, was to seize control of Lebanon. As far as Hezbollah (and other Arab regimes) was concerned, destroying Israel and replacing it with a Muslim state was secondary to securing their own survival in their respective countries. While the PLO might have given up its grip on Lebanon if all of Israel were captured, the Hezbollah's domination of Lebanon was due to its political and military might. Ironically, the Hezbollah needed Israel as an excuse for its military existence as a deterring force against Israel invading Lebanon. In fact, the strength of the Hezbollah, mostly its rockets and missiles, as with the PLO in 1982, increased the probability of another Israeli attack on Lebanon.

# 7

# Israel's Military Strategy and Doctrine and Its Resemblance to Western States

On 4 November 1948, Ben-Gurion advised the IDF to study theories and patterns of fighting of other militaries.[1] Indeed, over the years there has often been a resemblance or other linkage between Israel and other states (mostly Western ones) in terms of strategic and operational dimensions. Examining this subject will help readers understand the way in which Israel has coped with high- and low-intensity wars.

## The Threat to the Existence of the State

In February 1959, commenting on the defeat of France in 1940, Ben-Gurion proclaimed that the IDF could not afford such a failure.[2] By then Israel had lost battles—even important ones—during the 1948–1949 war, such as in the Negev during the first stage of that showdown. Britain also had its share of military setbacks at the beginning of World War II. Still, Israel and Britain managed to avoid the fate of France in 1940, which lost a high-intensity war. France had to give up territory (as well as part of its independence) and exposed many of its people to great suffering at the hands of Germany from 1940 to 1944.

In case of a total disaster in a high-intensity war, Israel faced a bigger catastrophe compared to France in 1940. In spite of the hostility between France and Germany at that time, the latter had not tried to obliterate the existence of its traditional foe, as it later wished to do to the Soviet Union in 1941, and as the Arabs wished to do to Israel if they had won. An Arab victory would put the Israelis in permanent danger of being neutralized or driven into exile.

France was liberated in 1944 by the Western Allies. Israel could not expect any state, including those that had close military relationships with it, to come to its rescue. Possibly the only hope for Israel would be the link between its survival and some vital interest of foreign powers (for example, Israel's relative proximity to the Suez Canal and oil fields in the Persian Gulf). Those strategic assets had worldwide importance during the Cold War, and particularly in a global showdown between Western states and the Soviet Union. France was worth fighting for because of it being a springboard to Germany, while Israel would be a jumping-off point to oil fields and the Suez Canal. If Israel had fallen to the Arabs during a war between the Soviet Union and Western powers, the latter might have considered retaking it, as they did with France in 1944. Like the people of France in 1944, those in Israel would have probably had to pay a heavy price during an attack launched by Western militaries on their occupied land. Yet many Israelis, similar to a lot of Frenchmen in 1944, would have been willing to absorb casualties in return for a chance to gain back their state.

"Preventive war is presented in advance as a prospective justified war."[3] States can find themselves facing a serious change to their status quo[4] that might happen in a matter of years, or even months. The assumption of preventive war is that attacking as soon as possible would be the best, if not the only, solution to the upcoming threat, instead of waiting for an uncertain future to unfold. Yet starting a high-intensity war could deteriorate into a complicated confrontation jeopardizing the infrastructure, cities and civilians of the country that began the hostilities. This was true in Israel's case too. Furthermore, its preparations for this kind of clash could have been discovered by the Arabs, who already suspected that Israel wished to expand at their expense. As a result, they might have increased their readiness or even tried to strike first. Ironically, this development could have convinced the Israeli political elite of the necessity of a preventive war.

The United States opposed the Israeli offensive on Egypt in 1956 and considered imposing sanctions against Israel in order to force it to withdraw from the areas the IDF conquered in Sinai. Ben-Gurion claimed that if the United States had stood in front of a clear and present danger to its existence, as Israel did at that time, the superpower would have adopted an approach similar to what Israel chose.[5] Indeed, preventive war was part of Western military thinking,[6] and from Israel's perspective the 1956 confrontation was a preemptive war. When the battles started in 1956 the Egyptian military was not yet ready, since it had not assimilated all the weapon systems received through its enormous arms deal with the Soviet Union. An Egyptian invasion was therefore unlikely in the upcoming months, but it could have happened in a few years. Israel believed it had to stop in advance an opponent who might become a threat in the future, as Germany in 1941 would have jeopardized the United States later on in the 1940s. Egypt also took other aggressive steps against Israel, such as blocking the Tiran Straits, which could have undermined Israel's economy in the long run.

On 29 January 1958, Ben-Gurion claimed that the 1948–1949 war was "one of the rare events in history," if not unprecedented, since that war was declared against a state on the day it was established, and its military was created during the battles.[7] Likewise, the United States had its war of independence. In both showdowns the odds were against the new states, since their foes enjoyed an obvious military advantage. Gradually, however, after a tremendous effort, both Israel and the United States managed to win their respective wars, albeit without forcing their enemies into a complete surrender. As a result, the two sides confronted each other again: Israel clashed with the Arabs in 1956, and the United States collided Britain in 1812.

B. H. Liddell Hart claimed that the goal of any war is to achieve a better peace.[8] In World War II the Soviet Union described its advance into Eastern Europe as one that liberated nations, but the control of countries like the Baltic States did not revert to the local population. Their rulers just changed.

Arab states since 1948 have publicly supported the idea of a Palestinian state. At least until the peace treaty between Israel and Egypt in the late 1970s, all the Arab states officially strove to replace Israel—considered by the Arabs a foreign entity in the Middle East—with a Palestinian state. Yet if Israel had been conquered by Arab states, it is doubtful that they would have handed over to the Palestinians the territory that was once Israel. From 1948 to 1967 the West Bank was an integral part of Jordan, while Egypt controlled the Gaza Strip. Those two areas, populated by hundreds of thousands of Palestinians, were not given to them. This fact pointed out the intentions of Arab states on this issue, demonstrating that a total Arab triumph over Israel would have not necessarily meant the fulfillment of the Palestinian national dream. It would have probably initiated another bitter struggle, only this time between Palestinians and Arab states. It is also probable that even if Arab states had allowed the establishment of a Palestinian state, they would have ensured their control over it, just as the Soviet Union did in Eastern Europe from the late 1940s. to the early 1990s.

The commitment by France and Britain to help Poland in 1939 did not prove a solid guarantee at the moment of truth when Germany invaded Poland.[9] In late October 1956 Israel reached an agreement with France and Britain against Egypt. Israel counted on those European powers to destroy the Egyptian air force threatening Israeli cities. After the war started, there was a delay in the offensive of France and Britain, which caused deep concerns within the Israeli leadership. Meanwhile, the Israeli rear was exposed to Egyptian air bombardment. For Israel, it was a valuable lesson: foreign powers could go back on their promises in times of war (although in 1956 they eventually fulfilled their obligations). Israel was aware that it must rely on its own forces, as Poland had to do in 1939.

## The Nature of a Long Struggle

In 1953, according to the Israeli intelligence branch, Arab chiefs of staff viewed the conflict with Israel as one that at the moment was in its "cold" stage but could change into a "hot" war.[10] This resembled the global "cold war" between the United States and the Soviet Union. In both cases there was a probability of a direct high-intensity war between the two sides. Israel had such a collision every decade until the 1980s, while the United States kept preparing for a scenario that never occurred. In both of those cold wars there were also proxy wars. In the mid-1950s Egypt used the Palestinians against Israel; the Soviet Union implemented the same strategy in countries like Vietnam.

After the 1967 showdown Israel was convinced that it had military superiority over Arab states.[11] Britain and the United States had been in a similar position regarding Germany after World War I; yet, like Israel in 1973, they, too, had to face the same enemy in the future after its forces were rebuilt. At least as far as Israel was concerned, the Arab catastrophe in 1967 made it clear that its foes had given up the "march to Tel Aviv" mentality. By this time, the Arabs had ceased to regard the military struggle as a short voyage ending with the total collapse of their rival.

After the 1973 showdown, some Arabs tended to believe that Israel was about to crumble. The Israelis were compared to the Crusaders, who were weakened from within before they collapsed completely.[12] Indeed, there was a certain resemblance between the Israelis and the Crusaders. Both of them were drawn—because of religious reasons—to the same territory, where they created a state that had to deal with the same foe, which had numerical superiority. Israel should have learned from the mistakes of the Crusaders, and only time will tell if Israel survives longer than the various Crusader kingdoms.

## Regime Change

In 1950, at the beginning of the Korean War, after the landing in Inchon, the tide of war turned in favor of the UN forces, which were mostly composed of U.S. troops. Senior figures in this camp, such as General Douglas MacArthur, thought to launch an offensive into North Korea and throw over the regime of Kim Il-Sung.[13] In 1991, during the showdown in Iraq, the United States wished that one of the results of the collision would be the downfall of the president of Iraq, Saddam Hussein.[14] Israel likewise hoped, in several wars, to get rid of the leader of Egypt—Gamal Abdel Nasser. On all those occasions, the leaders the United States and Israel sought to remove stayed in power in spite of their failures in combat. This demonstrated the limitations of using military force for regime change.

Ben-Gurion once claimed that there were unavoidable wars, such as those that erupt against a growing danger, as was the case in both World War II against Hitler and the war of 1956.[15] France paid a high price because it did not recognize in time Hitler's threat. In 1956, when France joined Israel to fight Nasser, the European power was well aware of the need to handle a dictator while he was still not too strong.[16] For the British prime minister in 1956, Anthony Eden, who suspected Nasser's intentions toward Western states[17] and who had also experienced World War II, Nasser was a kind of a Hitler who must be stopped. In the 1982 war the Israeli prime minister, Menachem Begin, despised the head of the PLO, Yasser Arafat, and saw him as a successor of Hitler. Like Israel, France and Britain in 1956, Begin strove to beat an Arab leader before he became a Hitler-like threat, even if this outcome was quite unlikely.

Another way to eliminate hostile leadership, including military figures, was assassination. On 29 October 1956, just before a war started, Israel sent a jet that intercepted an Egyptian plane that was supposed to have the chief of staff of the Egyptian military, General Abdel Hakim Amer, on board of it. He himself was not, but senior Egyptian officers who were on that plane were killed.[18] It was similar to the ambush that occurred on

13 April 1943 when U.S. fighters shot down the plane of the Japanese admiral, Isoroku Yamamoto; the death of this key military commander was a severe blow to Japan.

## Facing a Coalition

In 1956 Ben-Gurion was worried that just as Mustafa Kemal Ataturk had united the Turks, a dominant Arab figure would gather its people around him, and this step would be used against Israel,[19] exploiting the enormous advantages of the Arabs over Israel, such as the size of the population. Nasser could have been such a leader. Yet, in order to beat Israel, the Arabs were not necessarily required to find an Ataturk of their own. A supreme commander such as Eisenhower would suffice. This kind of high-ranking officer had to be familiar with the disagreements between the Arabs and still plan how to maximize their military potential as part of a coalition. Nasser had a military background, but he was not a general. Furthermore, many Arabs were suspicious of Nasser's ambitions, so it was necessary to choose a purely professional military person. There were attempts in that direction. For example, in the 1967 showdown an Egyptian general, Abdel Moneim Riad, was the commander of the Jordanian army, but he was far from being the Eisenhower the Arabs so badly needed, and their humiliating defeat proved it.

In the 1800s, Napoleon's foes learned they should combine their efforts and not allow him to strike at each of them separately.[20] In 1813 the anti-French coalition implemented this logic and managed to upgrade the coordination between its members up to a level needed to overcome Napoleon's army.[21] Arab states reached the same conclusion, especially after the 1967 showdown, when the IDF defeated them one after another. At the beginning of the 1973 war, Egypt and Syria succeeded in collaborating and executing a plan they had agreed upon in advance. Although the Arab militaries did not win a decisive victory, it was their best performance yet.

## The Time Factor

Liddell Hart stressed the importance of the time factor in war.[22] Napoleon always strove to win quickly.[23] In the beginning of World War I, both sides likewise wished to achieve early victories.[24] The IDF also emphasized the urgency of a fast decision. However, in the 1956 war, the militaries of France and Britain acted at a careful and slow pace, as though they were about to cross the English Channel to confront the German army once again.[25] But the Egyptian forces were far from presenting such a challenge. In contrast, Israel maneuvered with great speed in the Sinai desert, as had been done in the World War II campaigns in North Africa.

Egypt's military in the 1950s and 1960s assimilated Soviet weapon systems and studied Soviet military doctrine.[26] During World War II the Red Army reinforced its bridgeheads in a rapid pace. The best way to prevent such action was to launch a counterattack as soon as possible.[27] In the 1973 showdown Egypt crossed the Suez Canal. The IDF, knowing that the Egyptian army was influenced by Soviet methods, tried immediately to destroy the Egyptian bridgeheads.

## Offense

Liddell Hart also argued that attack was the only way to beat the foe.[28] In the beginning of World War I, Britain and France tried to win by using that approach.[29] Israel's military doctrine was also based on the offensive. Like Western powers, Israel sought to hold the initiative from the start of the war, thus gaining momentum that would lead to a decisive victory.

J. F. C. Fuller claimed that conquering land was essential in order to verify the enemy's capitulation.[30] The IDF was aware of that concept, which often worked on the operational-tactical level; yet even after capturing vast territory in the 1967 and 1973 showdowns, Israel did not manage to force its foe to surrender.

Fuller and Liddell Hart both called for a deep penetration of armor units in order to create chaos inside the enemy's ranks, ending in its collapse.[31] According to Martin Van Creveld, in the 1967 showdown the IDF outflanked Egyptian forces in Sinai, just as the U.S. army did to the Germans in northern France in 1944.[32] The Germans in 1944 and the Egyptians in 1967 tried to hold their ground but failed. Soon they found themselves in a hasty withdrawal and under pursuit from the air and land. Due to their extending flanks, U.S. and Israeli forces were exposed to counterattacks. This was a risk the United States in 1944 and Israel in 1967 had to take, and it paid off.

Brian Reid has argued that Fuller and Liddell Hart thought it better to cause the foe to crumble than to destroy him physically.[33] In the 1956 war the main aim of the IDF was to bring Egyptian units to collapse.[34] This was accomplished thanks to rapid advance and encirclements of enemy concentrations, a maneuver that also succeeded in 1967 and in many other campaigns, such as in World War II.

Logistics is a vital factor in any war.[35] In the 1973 showdown the IDF invaded the southern part of Syria and moved toward Damascus. The danger here was that this advance could have sunk "into Napoleonic snow."[36] Although there was a huge difference between the voyages of Napoleon in 1812 and that of the IDF in 1973, the penetration of the latter into Syria extended its logistical operations, as happened to the French emperor. Like Napoleon's army, the Israeli units could have fallen into distress because of a supply shortage, which would have exposed them to counterattacks.

The terrain has an enormous impact on military maneuvers.[37] Napoleon's army used physical features such as rivers or mountains to cover its flanks.[38] The IDF did the same in the Golan on 8 October 1973, when the Rokad River prevented Syrian attacks on Israeli troops.[39] Three days later, when Israel's 36th Division advanced toward Damascus, its northwest flank leaned against the slopes of the Mount Harmon.[40] Moreover, the IDF—fighting over and over again in the same areas, such as the Golan and Sinai—became familiar with those territories. The same thing happened to the British military, following its many battles over the centuries in some sectors in northwest Europe.

In 1956, when Israel, Britain and France planned to attack Egypt, France was up against an offensive in winter that could have disrupted its landing from the sea and from the air.[41] So the war started in late October instead. In 1982 Israel did not wish to postpone its offensive to the point that it would have been attacking in the relatively hard climate of Lebanon in the winter, which could have delayed its advance, if not put it on hold. Therefore the war began in early June, in comfortable weather.

The contradictions between the real Israeli goals in 1982 and the ones that were officially declared disrupted the Israeli offensive. The same happened with the attack of France and Britain in 1956, because of political deceptions that were part of it.[42] Indeed clarity is essential in an offensive. At the same time, there was a need to cause disarray, such as by breaking the foe's communication, which has been done—using various methods— since the days of capturing messengers riding a horse. Using the artillery to cut off communication lines was the next stage. In World War II, as in the 1973 war, jamming wireless communications marked the beginning of electronic warfare.

# *Defense*

The strategy of the Confederacy in 1864, which involved protecting all of its sectors, helped the Union Army, which enjoyed superiority in the number of both troops and war material.[43] In the 1973 showdown Israel also tried to cover every piece of land on its border, which played into the hands of its foes, particularly since the Arabs had many more soldiers and weapon systems. The Confederacy and Israel should have concentrated their troops in fewer sectors instead of overspreading them. This mistake made it easier for their respective rivals to exploit their quantity advantage. In 1864 and 1973, political considerations pressured the armies to use this kind of forward deployment, and the consequences in the battlefield were catastrophic in both wars.

In the 1960s there was a concern in the United States that the Soviet Union might commence a surprise attack in Europe. The prospective aims of the offensive ranged from conquering Western Europe to much more limited goals, such as capturing the city of Hamburg.[44] The IDF suspected in the early 1970s that Syria might start a full-scale offensive to regain all the Golan, or else just launch a much smaller operation to seize a tiny part of this territory.[45] Therefore, the United States and NATO in the 1960s and Israel in the 1970s had to be on alert for all kinds of scenarios.

In the 1970s, the battlefield between Israel and Arab states, because of the amount of troops and weapon systems that both sides possessed, appeared to be such that Israeli casualties could have been immense if there were a high-intensity war. One of the options of the IDF was to adopt a defensive approach: wearing down the Arab forces with massive firepower, mostly from the air.[46] This conception was similar to the one NATO had in Europe at the time.

On April 1956 the training branch of the IDF considered the way the British 8th Army stopped an Axis offensive in Alam Halfa in Egypt during August 1942 as a model for deployment against armor.[47] Such a move could have become relevant in 1956 if Egypt's military had attacked Israel after assimilating hundreds of armor vehicles.

The wrong impression the Israeli public got before the 1973 war was that the military posts near the Suez Canal were a constant line of defense like the Maginot Line in France in the 1930s.[48] But the Israeli strongholds were not necessarily built for a high-intensity war, in contrast to the Maginot Line. Perhaps the most common aspect the two defense lines shared was their being both associated with military failures that occurred when the war broke out.

Moshe Dayan, as the minister of defense in the 1973 war, admitted that Israel had setbacks, such as the loss of the line of posts near the Suez Canal. However, according to him, it "should be remembered what happens at war." He maintained that events like Dunkirk, the fall of the Maginot Line, Pearl Harbor, and so on "must give us a realistic perspective."[49] Similar to the United States in Pearl Harbor, Israel in 1973 absorbed a surprise attack but subsequently managed to turn the tide in its favor.

## Air Power

World War I marked the beginning of modern warfare.[50] This included the appearance of the fighter-bomber, since it could drop bombs and intercept planes with its machine guns. This weapon system has been significantly upgraded ever since its first appearance. The IAF, since its establishment, has been based on the fighter-bomber.

In 1951, a research study conducted by the planning branch of the IDF examined ways of achieving air superiority, based on the lessons of World War II.[51] Even before that, on 3 November 1949, Ben-Gurion, as the prime minister and minister of defense, claimed that the lessons of both World War II and the 1948–1949 war proved "without any doubt" that an air force is essential in winning a war.[52] Contrary to World War II, in 1948–1949 the plane did not play a major role, but Ben-Gurion's conclusion was verified in future showdowns.

After World War I, the British air force had to compete for budgets with the other branches of the British military. This was done at a time when another high-intensity war was not expected.[53] After the 1948–1949 war, in a relatively calm period, the IAF had to compete with other branches in the IDF to obtain funds.[54] The IDF was created in 1948; yet its ground units existed as part of the underground decades before the air force was established. The IAF, like the British air force in the 1920s, needed to deal with land forces that had seniority over it.

Air bombardments were essential in the war in North Africa in 1941, when British planes disrupted the effort of the Axis Powers to push supplies and reinforcements to their troops.[55] The IAF did the same in Sinai in the 1956 and 1967 wars. On other occasions, such as sending German land units to Anzio in Italy during 1944, or when Iraqi convoys flowed into Syria in the 1973 showdown, air power did not block their arrival at the front line. In both cases the ground forces, traveling hundreds of kilometers, were exposed; yet they reached their destinations.

Gunships were supposed to balance the quantity inferiority of NATO in Europe.[56] The IDF needed heavily armed attack helicopters for a similar reason, especially when facing a surprise Arab offensive. The accurate firepower and availability of the choppers were essential in stopping an enemy offensive.[57]

During the Vietnam War the anti-aircraft missiles were used to protect ground maneuvers, such as in March 1972.[58] This was part of the Soviet military thinking,[59] which influenced Egypt's military.[60] At the end of 1969 Nasser believed the IAF had attacked the anti-aircraft missiles in order to stop Egypt's military from crossing the Suez Canal.[61] At the beginning of the 1973 war, however, the IAF did not manage to prevent the Egyptian anti-aircraft batteries from securing the offensive of the Egyptian forces when they seized the east bank of the Suez Canal.[62]

## Tanks Versus the Infantry

In its early stages, the Roman army was a militia trying to avoid structure problems by including many veterans in its ranks.[63] The IDF, which also starting as a militia, did the same. Rome and Israel (at least in the years 1948–1956) also based their militaries on the infantry. Those two militaries operated in different eras, but they still counted on the infantry. The tactics and certainly the weapons have changed, but the basic formation of the army was the same. Another veteran corps that played a central role was the cavalry, which was replaced in the 20th century by tanks. This raised a dilemma: Which corps should have the upper hand?

In the 1920s there were serious doubts within the British general staff regarding the tank,[64] just as existed in the IDF until the 1956 war. For high-ranking British officers, World War I did not provide proof of the tank's dominance. In the 1950s senior Israeli officers thought the same— that is, they underestimated the tank in spite of all that was known about it following two world wars.

On 13 February 1959, Ben-Gurion claimed that the French military was beaten in World War II because it was prepared to fight a war like World War I.[65] It was a well-known lesson. Still, in the first half of the 1950s, when Ben-Gurion was the minister of defense, the IDF assumed its next war would be like the former showdown in 1948–1949, when the infantry was the main corps. However, although the IDF prevailed in the 1956 war, it went through a drastic change afterward, beginning to rely on the armor instead of the infantry, a move long overdue.

In World War II, a crew of gunners with a 50mm cannon could have destroyed a tank.[66] Others used rocket launchers that also could stop a tank, just as the English longbow knocked out the French cavalry in the Battle of Agincourt in 1415. In 1973 Arab infantry could have neutralized a tank with anti-tank missile, a quite new weapon then. The rest of the anti-tank arsenal (rockets, artillery, etc.) appeared already in World War II. The anti-tank missile was more accurate and sophisticated than other anti-tank weapons; yet—as in World War II—in the 1973 showdown it was the combination of anti-tank measures that broke the armor assault. In 1973, the Israeli armor, which already tended to storm its foe in almost every scenario, fell into an Egyptian trap. The vulnerability of the tank to enemy fire caused the Israeli armor heavy casualties.

Some have argued that worldwide the tank had started its decline in 1967[67] or in 1973,[68] when the IDF went to war trusting in its armor, owing to armor's success in the last showdown in 1967. Israel did not lose in 1973 because of this approach; yet Israel's overdependence on the armor was clear, and in a way this war rocked the IDF to its very foundations. However, in spite of its shortcomings, the tank remained the backbone of the Israeli ground forces after the 1973 war, since the tank had proved itself in all kinds of scenarios. Indeed, the Arabs did not beat the IDF because of the Israeli investment in the armor—that is, increasing the number of tanks, upgrading them, training their crews, and so forth. Those preparations were not perfect, but they were enough to both prevent a complete catastrophe and obtain important achievements.

## Combined Arms

In the era of Napoleon, artillery, infantry and cavalry had to assist each other,[69] with the cavalry covering the flanks of the infantry and vice versa. There were times in which one of the corps was considered dominant up to the point that it could encounter the enemy alone. Yet there was an obvious need to create a formula balancing the weaknesses and advantages of both corps. This applied to joint operations of land, sea and air as well.

The importance of combined arms in the British and U.S. militaries was demonstrated in World War II.[70] In 1951 the planning branch of the

IDF emphasized the mutual dependence between land, air and sea forces. It was recognized that planning and executing combined operations at the end of World War II had resulted in "a great success."[71] Accordingly, the IDF tried to train its forces in combined arms during the 1950s and 1960s.[72] In the 1970s such cooperation was called "the integrated battlefield" by the IDF. An Israeli general acknowledged that it was not a new idea per se, since it was implemented in World War II by Western militaries.[73]

In 2012, in spite of American superiority in air and sea combat, "relying solely on long-range precision platforms for deterrence greatly reduced the strategic and operational options available to our national policymakers." Ground forces were needed as well.[74] This approach was also relevant to Israel. In 2013 the IDF tended to depend on accurate firepower, mostly from the air, while reducing the role of the land units. Both the American and Israeli militaries had to maintain their ability to conduct ground maneuvers.

Aside from the differences between the arms and the corps, there was also a distinction between units inside the corps—for example, in the infantry—which required cooperation between them. The slow rate of fire and limited range of muskets in the 17th century exposed the troops using them to an enemy attack when reloading. Therefore, another category of infantry, one that was equipped with spears, protected the musketeers. In the 20th century, in militaries such as the IDF, there was cooperation between various categories of the infantry too. When part of the infantry force was storming the enemy, another part was pinning it down with mortars, heavy machine guns, and so forth.

## Naval Warfare

Israel resembles Britain in being an island. Whereas the latter is surrounded by water, Israel is surrounded in the Middle East by hostile Arabs. During a war, similar to the British navy, the Israeli navy must secure its country's sea routes. In Israel's case, this meant lanes in the Mediterranean Sea (among others). It was essential to maintain the connection with friendly states in Europe. In the 1973 war, during a series of naval skirmishes, Israel defeated the fleets of Syria and Egypt, thus keeping Israel's sea routes open. The Arabs who declared a blockade on Israel could not enforce it and cut Israel off from its supply sources, a situation Britain almost came to in certain periods during World War II.

Naval history shows how vessels of the same type clashed with each other at sea. The fleets of Israel, Egypt and Syria in the 1973 war were based on missile boats. As in a battle between aircraft carriers in World War II, the missile boats in 1973 sank their foes without needing to see them, just by launching anti-ship missiles instead of planes with bombs or torpedoes.

Liddell Hart criticized landing operations in World War II such as the one that took place in Anzio, since they were not exploited properly, as were amphibious assaults in the Pacific Ocean.[75] The IDF, after landing its troops on the shore of Lebanon in the 1982 war, made the same mistake, since its units were delayed too much in the beachhead instead of breaking inland. In other wars Israel's ground forces assisted the navy by seizing naval ports, as Western militaries did in France and Holland during 1944.

## Low-Intensity War—Strategic Aspects

Since its establishment, the United States has been involved in less than 12 conventional collisions. Most of its wars were asymmetric ones.[76] In a way, the same could be said about Israel. Since 1948 it has had five high-intensity wars together with numerous skirmishes and incidents that had to do with asymmetric warfare, from Palestinian infiltrations in the early 1950s to dealing with the Hamas and Hezbollah in the last decade.

Daniel Byman, research director of the Saban Center for Middle East Policy, argued in 2011 that Israel's achievements and setbacks in fighting terror "can serve as a blueprint for all countries fighting terrorism." Many states learned from Israel, which could teach them much more.[77] Of course, those states must be aware of the differences between their situations and that of Israel, just as Israel has to adjust lessons learned from others in this field to its own conditions.

On 11 September 2001 the United States lost almost 3,000 people, nearly all of them civilians. In response, its military moved the battlefield into enemy land (Afghanistan) and seized that country. In the 2000–2005 confrontation between Israel and the Palestinians, more than 1,000 Israelis were killed, many of them civilians. The peak in those deadly assaults, in March 2002, caused Israel to transfer the fight into enemy territory in the West Bank and occupy that area.

Al-Qaeda did not predict the American response in Afghanistan after 9/11.[78] In Israel's case there were many occasions on which its foes in hybrid and low-intensity wars were surprised by its powerful and swift reaction. Examples of this can be found in Israel's attacks on the Hezbollah in 2006 and on the Hamas in 2008–2009 and November 2012.

As with Japan and Germany after World War II, the United States wished to turn Afghanistan from a foe into an ally and have military bases there against common enemies like the Taliban. Israel has been trying to do the same with the PA. IDF's camps in the West Bank are set up against Israel and the PA's mutual enemies, mostly the Hamas. The Afghan government and the PA are under threat from a foe that seeks to overthrow them and impose its strict Islamic laws on their populations.

The Afghan government and the PA don't control all the territory that officially belongs to them. In fact, they might lose even the area they hold. The United States is worried that after its withdrawal, the Taliban, if not Al-Qaeda, might return to power in Afghanistan. The superpower strongly opposes going back into this hornet's nest. Israel has a similar concern about zones it left in the West Bank, along with possible additional territories presented to the PA as part of a future agreement. Israel had already retreated from these areas following the Oslo peace accord in the 1990s. However, guerrilla and terror assaults originating from those zones forced the IDF to reoccupy them, in the 2000–2005 confrontation. Israel does not want to repeat this painful process.

For several years since 2003, insurgents in Iraq have destabilized the pro-American regime there, which exposed the weak spots of United States' conventional superiority.[79] The rebels got assistance from Syria and Iran.[80] Neither country was held accountable, although the United States could have attacked Syria and/or Iran. Since the 1990s, Israel, in its hybrid/low-intensity wars in Lebanon, the West Bank and the Gaza Strip, has also faced (albeit indirectly) Iran and/or Syria. Israel also could have confronted those states head on but concluded—just as the United States did following the Iranian and Syrian involvement in the low-intensity war in Iraq—that undertaking such a collision might not be worth it.

In the war in Afghanistan, Pakistan was supposed to be a U.S. ally. Nevertheless, insurgents, weapons and ammunition flowed from Pakistan to the U.S. enemies in Afghanistan. Israel faced a similar problem. Guerrilla and terror fighters, along with war material, were smuggled into the Gaza Strip from Sinai, an Egyptian territory. Israel has a peace agreement with Egypt, but the latter failed in blocking its border, particularly until 2013.

In the 1950s, France in Algeria, Britain in Malaya and Israel in the West Bank and the Gaza Strip had to conduct low-intensity wars. While for France in Algeria, and especially for Britain in Malaya, the front line was far away, in Israel's case it was much closer. This constraint also had an advantage for Israel, since it shortened its supply lines. Yet the strength of France and Britain gave them no less power and influence in their remote colonies than Israel had in the areas nearby it. This was no surprise, since they were European powers, while Israel in the 1950s was a new, small and relatively weak state.

For the United States the wars in Iraq and Afghanistan were low-intensity wars, while an outburst in Korea would have been a high-intensity war. Israel might have faced a similar challenge if during a low-intensity war in the West Bank the IDF had to send its forces to confront Egypt or Syria in a high-intensity war.

## Low-Intensity War—Operational Aspects and Buildup

The area where Jews lived prior to the establishment of modern-day Israel was called the Yishuv. At that time it was a territory under Ottoman rule and subsequently a British Mandate. Security problems haunted the Jews of the Yishuv, but they were forbidden to form their own police, let alone a military, to protect their lives and property, since the foreign authorities feared those forces could be turned against them. The Jews had only local militias under their own control, and at best (at least until 1948) underground forces such as the "Hagana." All those groups were lightly armed, and most of them were quite small, poorly trained and basically suited for tasks such as defending their local town or kibbutz.

Israel to a large extent based its legal approach in low-intensity war on British procedures from the period of the Mandate.[81] Furthermore, during the Arab rebellion in the late 1930s a unique unit was created: the "Special Night Squads," which included Jews from the Yishuv and British troops. Together they went to battle against their common enemy.[82] During the 1930s and 1940s Jews from the Yishuv joined security and military services that were subordinated to Britain. These Jews fought not only for Britain but also against the Kingdom, particularly in 1945–1948. Thus, by 1948 the Israelis had diverse experience in low-intensity warfare after clashing with or against a foreign rule.

The militaries of Israel, Britain and France adjusted to confronting guerrilla and terror by giving freedom of action to their officers, including those in junior ranks.[83] This approach was much needed, since sometimes senior officers tended to be overinvolved in an operation. They wished to help, but this could have brought junior officers to rely too much on receiving orders instead of solving problems on their own.

Along with the Israeli experience in the wars in Lebanon in 2006 and in the Gaza Strip in 2008–2009, "the American experience in Iraq in 2003 is a useful case study of combat between a conventional western military and an adversary attempting to fight as a hybrid threat."[84] In 2000 the U.S. Army realized that urban warfare should be part of standard exercises, like desert training, and not a task the military should be coping with only as a last resort. Later on, in Iraq, the American military gained experience in urban warfare.[85] In past low-intensity wars—such as Vietnam—it was not that common, except in battles like in the city of Hue in 1968. The IDF also experienced low-intensity wars that did not require a lot of knowledge in this field, such as those in the 1950s and the 1960s, while in the confrontation of 2000–2005 urban warfare was much more dominant.

In Vietnam the enemies of the U.S. military hid in underground tunnels and bunkers.[86] Palestinians in the Gaza Strip did so as well during their first mutiny against Israel in the early 1970s.[87] Since 2000, Palestinians in the same area have expanded the use of tunnel warfare to move supplies, men, and so on, from one sector to another, including to countries nearby: Israel and Egypt.

In the 2006 war in Lebanon, "although knowledge of conventional warfighting still existed in the IDF, the lack of training in that area created a shortage of military expertise in conventional operations. Israeli ground forces found themselves in a tactical disadvantage when fighting Hezbollah in a hybrid war. Swinging the U.S. Army training pendulum too far to the right (the conventional paradigm) or too far to the left (the irregular paradigm) has the potential to create similar issues for future U.S. Army forces engaging in hybrid warfare."[88]

Rome—in its time—used targeted assassination against leaders of guerrilla army, just as the United States has done against Al-Qaeda and as Israel has done against Hamas.[89] There was obviously a difference in the tactics between the ages, but the method was similar: to capture the target by means of surprise. It was not easy, since guerrilla leaders don't conduct their operations similarly to a conventional military—that is, with a big headquarters, particularly one placed in the same site for a long time. Instead, these commanders live on the run, watching their surroundings, knowing that their superior foe is trying to exploit its edge in order to capture or kill them. Indeed, guerrilla and terror armies, because of their improvised and nonprofessional nature, are often based to a large extent on one charismatic and skillful figure. Neutralizing that figure could inflict a major blow to the morale and performance of the fighters.

The U.S. Army has transformed into a force focusing on COIN (counterinsurgency) missions. Accordingly, its structure shifted in 2013 to rely on light infantry instead of armor units.[90] The IDF has in recent years made changes in order to be better prepared to run hybrid and low-intensity wars. One shift was increasing the number of infantry in armor battalions.

## The Challenges of the Afghan and Palestinian Security Forces

One of the major American concerns in Afghanistan was the ability of the Afghan security forces to accomplish their missions after the American withdrawal. Israel has similar doubts about the ability of the Palestinian security forces to hold their ground without the IDF's support.

For a few decades now Afghanistan and the West Bank have served as battlefields. Yet their security forces are not born soldiers. They might be accustomed to combat, but they had to learn the art of war in a professional way, which required them to study Western powers. Improving military education in Afghanistan was essential in building its security force,[91] as was keeping politics out of the training of Palestinian security forces.

Although the Afghan forces since 2001 and the Palestinian security forces following the 2000–2005 confrontation had to be built from scratch, they had some advantages. First of all, both entities were familiar with their land and foes, sometimes even personally, which enabled them to communicate with the local population. Furthermore, fighting on their own soil boosted their motivation and contributed to their will to demonstrate courage and devotion.

Afghan and Palestinian security forces don't have to handle high-intensity wars against foreign militaries. All they need to do is manage low-intensity warfare against non-state organizations. The Afghan security forces' main enemy has been the Taliban, while Palestinian security forces confronted the Hamas. Afghan and Palestinian security forces should be qualified to enforce law and order and carry out military missions such as limited assaults. Many of their tasks are even far less demanding, such as manning roadblocks, going out on patrols and fulfilling various guard duties in camps and key civilian facilities. Therefore, Afghan and Palestinian security forces don't require weapon systems like bombers, heavy artillery, tanks, and so forth. The armor vehicles they already possess should be sufficient for their work.

The United States provides money and instructors to Afghan and Palestinian security forces, although there has been a concern that they would exploit their knowledge and weapons against Israeli and American objectives.[92] For example, there have been incidents in which Afghan troops have murdered American soldiers. Israel also absorbed casualties from Palestinian security forces in the 2000–2005 confrontation.

## The Population as a Factor in a Low-Intensity War

Moshe Dayan, eight years after his term as chief of the general staff, visited Vietnam in the summer of 1966. He claimed that foreigners—that is, the United States—were not capable of administering that country for the long run. According to him, their mistake was thinking it was possible and assuming the local population would be grateful to them. Dayan, as minister of defense in Israel after the capture of the West Bank in the 1967 war, strove to avoid unnecessary involvement in the lives of the Palestinians if there was no threat to an Israeli interest.[93] Dayan thus tried to avoid repeating the error—according to him—that the United States had made in Vietnam. However, Israel supported progress—such as improving education—in the West Bank, believing the Palestinians would approve of it. By doing so, Israel somehow followed the pattern the United States established in Vietnam. The 1987–1993 Palestinian uprising proved that was a wrong step.

Yitzhak Rabin argued that in Israel's war in Lebanon against the PLO in 1982, Israel could not avoid hurting the Arab population there, because of "the way the terrorists confronted" the IDF. According to Rabin, the IDF was far less brutal and took better care of civilians than Western militaries did in wars such as Algeria and Vietnam. Still, Rabin added that this kind of warfare, in which Israeli troops coped with noncombatants and fighters, "left scars" and caused unease among many Israeli soldiers.[94] Rabin, who served as prime minister and before that as chief of the general staff, did not play an active role in the war of 1982. Yet, like some former senior leaders in Western states, he claimed that his state's military had abided by high moral standards. Those dilemmas appeared in other wars, such as the conflicts in Afghanistan and in the Gaza Strip in 2008–2009. "Like Hamas in Gaza, the Taliban in southern Afghanistan are masters at shielding themselves behind the civilian population."[95]

In Vietnam, during the battles of 1968, American firepower caused heavy casualties among the population and severe damages in urban areas such as Hue. Following those events (as well as the political downfall of the South Vietnamese government), the United States military had to reconsider its tactics.[96] The IDF had similar concerns in bombing targets of the PLO in Beirut in 1982 and during the struggle with the Palestinians in the West Bank and the Gaza Strip from 2000 onward.

Since 2001 the United States has poured hundreds of billions of dollars into the war in Afghanistan. Since 1967, when the West Bank was seized, Israel has allocated billions of dollars for this cause. A large part of the American and Israeli budgets in those places was destined for civilian purposes, like building and improving infrastructure. In Israel's case, those projects were mostly for the benefit of its Jewish people, not for the Palestinians, while the United States' effort—not having any settlements in Afghanistan—was for the local population. For both the United States and Israel, assisting the population was part of the overall strategy of holding their ground. Both countries found themselves investing huge amounts of money in areas that were often battlefields. Furthermore, the United States planned to leave Afghanistan, which put at risk its entire investment in that war-torn and poverty-stricken country. It could have been the same for Israel if it had left the West Bank.

## Nuclear and Chemical Warfare

Chemical weapons first appeared some 4,000 years ago.[97] Egypt threw gas on its foe during the war in Yemen in the 1960s.[98] However, Egypt—even in its darkest hours in the 1973 showdown—like both sides in World War II during drastic situations, avoided dropping chemical weapons on its rival. The main reason, as far as Egypt was concerned, was the same as in World War II: the fear of massive retaliation.

According to non–Israeli sources, Israel has built a nuclear arsenal. Since this occurred only a few years after the nuclear age had begun, in contrast to thousands of years of low- and high-intensity wars, there was far less foreign knowledge Israel could have learned from in this matter. Therefore, Israel had to develop its strategy and doctrine on this issue more or less at the same time that it was done in other states in possession of this devastating arsenal.

In the 1973 war Israel might have been prepared to fire nuclear weapons if its conventional forces had collapsed.[99] This kind of scenario was always a possibility, considering the enormous advantages of the Arabs and their superior number of troops. This was also the case in Europe, where the overwhelming number of members of the Warsaw Pact might have pushed NATO to launch a nuclear strike if Western militaries were on the verge of defeat.

# 8

# The U.S.-Vietnam War and the Israeli-Hezbollah War in the 1990s

"The Vietnam War remains a reference point by which Americans weigh and evaluate overseas military intervention."[1] The Israeli fight in Lebanon against the Hezbollah in the 1990s was considered "the Vietnam War of Israel."[2] In spite of the huge differences between those two wars, they shared some aspects in terms of strategy, combat doctrine, and so forth.

## The Foe

In Vietnam the United States collided with the Communist government of North Vietnam. In Lebanon during the 1990s Israel encountered the Hezbollah, which was an organization, not a state, and should be compared to the Viet Cong rather than to North Vietnam. However, the Hezbollah was a sort of state inside a state, due to its influence and the major weaknesses of the regime in Lebanon. The Hezbollah strove to unite all of Lebanon under its rule and kick out Israel, just as North Vietnam hoped to do (and in fact did do) with regard to all of Vietnam while driving out the United States.

In the 1990s Israel struggled in Lebanon against an opponent that was familiar with the terrain, similar to the Viet Cong, whose men used their knowledge of the battleground in order to attack, hide, patrol, infiltrate and stealthily lay down ambushes and various types of explosive devises and deadly traps. In both cases, the locals' motivation was high since they were fighting for and on their land. They managed to penetrate and maintain a presence in areas that were supposed to be under the control of the United States/Israel.

There was a difference in military culture between Israel and the Hezbollah, as was also the case between the United States and its foe in Vietnam. The contrast was not the same in both wars; yet it showed a collision between Western and Eastern perspectives and how a democracy deals militarily with a non-democratic enemy.

North Vietnam and the Hezbollah both enjoyed external support, including military aid. The Soviet Union helped Vietnam as part of the global Cold War between the Soviet Union and the United States. Iran and Syria similarly stood behind the Hezbollah in Lebanon as part of their own cold war with Israel. Furthermore, during the 1960s the United States was concerned that a collapse of South Vietnam would push more states in Southeast Asia over to the Communist bloc. Likewise, Israel was worried a Hezbollah victory in Lebanon would encourage radical Islam in the Middle East, thus prompting Iran to continue casting its influence on other states. Such a possible development, known from the time of the Vietnam War as "the domino theory," urged Israel to continue fighting in Lebanon in spite of the cost and the risks.

The U.S. Navy did not try to intercept Soviet ships sailing to North Vietnam. Similarly, Israel had to accept the fact that supplies from Iran and Syria reached the Hezbollah in Lebanon in the 1990s. Israel, like the United States before it, wished to prevent an official conflict with another power. Even when force was used, the United States in Vietnam was only able to reduce the flow of supplies and reinforcements from North Vietnam to South Vietnam. Israel's military and security arms were not very successful in stopping the transfer of Hezbollah's troops, weapons and other war material to south Lebanon, either.

Syria used the Hezbollah in the 1990s to launch attacks on the IDF in Lebanon as part of the pressure on Israel during its negotiations with Syria about the Golan Heights. For North Vietnam, military actions during talks with the United States were meant to ultimately convince the United States to give up South Vietnam.

## Strategic Aspects

North Vietnam could have focused on the United States, just as the Hezbollah did with Israel in the 1990s, while the United States and Israel were busy on other fronts. Still, both the United States and Israel had an overwhelming advantage over their opponents, as expressed in the amount of troops, weapon systems and military technology they could have mustered. However, in the end both used a relatively small percentage of their military potential. In spite of deploying drafted soldiers, the majority of U.S. and Israeli manpower was not called up, which meant that on the battlefield the United States and Israel could not always capitalize on their obvious edge. This approach demonstrates that, despite all the importance of an American victory in Vietnam, and an Israeli one in Lebanon, both Israel and the United States did not consider those wars collisions that demanded a maximum effort. No wonder the United States in Vietnam, and Israel in Lebanon, never officially declared those confrontations to be wars.

The war in Vietnam proved that the ultimate goal of a war is to influence the enemy's will,[3] a well-known aim, which was also true in the collision between Israel and Hezbollah in Lebanon during the 1990s. Generally speaking, the United States in Vietnam, like Israel in Lebanon, wished to convince the adversary (whoever that may have been) to cease its attempts to attack them and their allies. The purpose was not the complete destruction of the enemy, but rather inflicting defeats upon it while preventing its troops from seizing key areas. Israel did not wish to possess any part of Lebanon, and the United States did not seek to annex Vietnam. Lebanon was for Israel what Vietnam was for the United States: a battlefield they would have been happy to leave, provided it did not fall into the hands of their enemy. Therefore, the presence of United States in Vietnam, and that of Israel in Lebanon, was a military one, without civilian settlements. This factor made it easier to handle the war, as well as the retreat.

Vietnam, like Lebanon, is quite a narrow country with a long shoreline, which allowed the United States in Vietnam, and Israel in Lebanon, to attack from the sea.

Furthermore, Hanoi, the capital of North Vietnam, and Beirut, the capital of Lebanon and the location of the Hezbollah headquarters, were exposed—being sea ports—to amphibious operations. But the United States avoided any major invasion of North Vietnam by sea or land, since this might have brought China or the Soviet Union into the war. Israel did not try such a maneuver, either, if only to prevent an escalation that could have forced Syria to intervene, possibly with the backing of Iran. The lack of a main offensive deep into the territory of the enemy was one of the main claims, within the United States and Israel, that the countries were fighting "with one hand tied behind its back."

In the 1991 Gulf War, one of the considerations preventing the United States and its allies from proceeding deep into Iraq was the need to avoid an unnecessary stay.[4] This was one of the lessons of the previous war in Vietnam, in which the United States had found itself entangled for an extended period, contrary to its intentions. Israel, too, in its collision in Lebanon in 2006, was very careful not to go back to its situation throughout the 1990s—that is, sinking into a quagmire of an ongoing struggle.

## Combat Doctrine and Operational Aspects

In Vietnam, the United States' method for estimating its achievements was by counting the numbers of enemy dead.[5] However, the stubbornness, devotion and courage of the Vietnamese made them willing to absorb heavy casualties.[6] This was also the case with the Hezbollah in Lebanon.[7] However, North Vietnam and the Hezbollah were well aware of their opponents' sensitivity to casualties, and how such losses undermined the morale of the enemy soldiers and society. The United States and Israel tried to protect their soldiers in various ways, such as giving them body armor.

The United States in Vietnam and Israel in Lebanon during the 1990s relied consistently on air power. North Vietnam had planes, while the Hezbollah did not, but both of them were very exposed to bombardments. The United States and Israel exploited this weakness in day-to-day activity, and also in major operations. Sometimes even an American/Israeli unit, caught in an ambush in an open area, could still be saved and ultimately win the battle, thanks to the availability and the lethality of airpower. Still, in both wars, due to limitations such as the conditions of the terrain, airpower was not always that effective against hostile infantry.

Israel, while in Lebanon during the 1990s, used the AH-64A gunship, which was built based on the lessons of Vietnam, as well as the AH-1 gunship, which appeared in combat for the first time in the Vietnam War.

Armor proved essential in Lebanon in the 1990s,[8] and sometimes also in Vietnam. The tanks in both wars could have used their firepower, mobility and protection against an enemy relying on infantry in ambushes and major operations, but the jungles of Vietnam and the thick vegetation in south Lebanon provided the enemy troops with hiding places and also made it difficult (and sometimes impossible) to deploy armor there. Often even the U.S./Israeli infantry found it difficult to cross those areas and spot their enemies on time.

The U.S. military in the Vietnam War and the IDF in Lebanon during the 1990s were huge organizations. They were also bureaucratic and complicated formations, which meant they needed time to adapt to an evasive and determined foe that relied on guerrilla and terror tactics. This resulted in repetitive patterns during the war in Lebanon, similar to what happened to the United States in Vietnam.[9] On the battlefield itself, the United States in Vietnam and Israel in Lebanon depended on troops from the regular service and standing army. Some officers and noncommissioned officers of the U.S. military and the IDF, however, returned to the same front again and again. The experience they gained every time was valuable for the next tour, but this routine was also a physical and mental burden. On the other hand, the regular soldiers in Vietnam and in Lebanon during the 1990s were deployed for a limited amount of time, after which they were replaced by others. There were, therefore, always fresh—albeit inexperienced—troops that had to adjust to the front, a process that took time, had its difficulties and sometimes cost lives.

During the Vietnam War the U.S. Army created facilities for learning and training against subversion warfare[10]; for example, before being sent to Vietnam, American airborne troops trained in a place that was built to look like a village in Vietnam.[11] In the late 1990s, Israeli soldiers underwent similar preparations before they entered Lebanon.[12] The purpose was, as in Vietnam, to get them accustomed to the local environment.

Guerrilla tactics were adapted by United States in Vietnam[13] and by Israel in Lebanon during the 1990s.[14] Although guerrilla warfare was known to be the strategy of the weaker side, it also suited the United States and Israel when they encountered a hostile guerrilla. This meant, for example, sending small groups of well-trained troops to infiltrate enemy territory in order to gather intelligence and lay down ambushes.

The United States in Vietnam, and Israel in Lebanon in the 1990s, had to deal with logistical challenges. Although Vietnam was about 13,000 kilometers away from the United States, while Israel shares a border with Lebanon, the arrival of troops, weapon systems and various war materials in both cases was basically a matter of organizing the transportation, fuel, and so forth. However, sending reinforcements and supplies to camps around Vietnam or inside south Lebanon in the 1990s, especially to isolated posts, was a different matter, due to enemy ambushes and physical features. This factor pushed the United States and Israel to dispatch helicopters for quick deliveries. The ability of the helicopters to hover and land in very small spots was essential in the battlefields of Vietnam and Lebanon during all types of missions, not only logistical ones.

## High- and Low-Intensity Warfare

The deployment of the United States military in Vietnam, and that of the IDF in Lebanon in the 1990s, provided those armed forces with combat experience. Their soldiers got tougher and were tested under fire in all kinds of terrain and weather. Troops were examined, from the privates up to the high command, in the fields of operation, logistics, intelligence, and so forth. The tours in Vietnam and in Lebanon became a kind of unofficial, yet necessary, stage in the career of officers and noncommissioned officers.

While the United States spent its resources on Vietnam, the Soviet military got stronger in terms of both weapon systems and numbers.[15] The amount of time, men, funds and equipment the United States had to invest in Vietnam was at the expense of preparedness for a high-intensity war. Such a collision could have happened because of Vietnam or as a result of another cause relating to the European theater of operations. During the 1960s the Warsaw Pact could have taken advantage of the fact that hundreds of thousands of American soldiers were pinned down in Vietnam and launched a major offensive in Europe. In such a case, the United States might have received an alert in time that would have allowed it to dispatch units from Vietnam before the showdown started in Europe. However, its troops would have needed time to accustom themselves to the new combat conditions. Those who had run the gauntlet of Vietnam could have found it easier to adjust to another clash, this time in Europe, but their Vietnam experience could have also been an obstacle.

There was some resemblance on the tactical level—that is, fighting a foe that possessed the same weapons in a similar landscape. After all, a jungle in Vietnam may have had something in common with a thick forest in Germany. Nevertheless, there was a major difference in many other areas. The Soviet military had some influence on the military of North Vietnam, but in Europe the United States would have met the Soviet troops directly and on a vast scale. American troops had to be prepared, physically and mentally, for scenarios that were rare in Vietnam (such as tanks versus tanks) if they wished to avoid a defeat and a heavy cost in both lives and weapon systems. The difficulties the U.S. military experienced after the Vietnam War emphasized the possible ramifications of that confrontation in case of a high-intensity war in the 1970s with the Communist bloc. Although the United States had nuclear weapons, they were kept as a last resort, because of the fear of an escalation to an Armageddon situation. Therefore, if the U.S. military had not gotten ready to match the Warsaw Pact in a high-intensity war, NATO, relying on the U.S. military, would have had to consider asking for a ceasefire according to Soviet terms, or else risk an exchange of unconventional punches on cities of both sides.

As far as Israel was concerned, dealing with organizations such as the Hezbollah also came, to some extent, at the expense of preparations for a high-intensity war.[16] Israel in the 1990s therefore had challenges similar to those of the United States in the 1960s and 1970s. Israel could have faced war with another power—Syria, for example—because of the clashes in Lebanon, or due to an outburst on another front (such as the Golan). With all the assistance Syria gave to the Hezbollah in its fight against the IDF, a direct and full-scale confrontation between the IDF and the Syrian military would have been a very different war. Like the U.S. military in Europe, the IDF had to be ready for armor warfare and giant collisions up to the level of corps, and not ones like in Vietnam and Lebanon, which mostly occurred on the level of battalions, companies and platoons.

## Local Allies

In Vietnam the United States hoped to create "a non Communist South Vietnam, free to choose its own form of government."[17] This included building up its military. Notwithstanding the importance of winning the war, the United States in Vietnam and Israel in Lebanon strove to limit as much as possible the need to risk the lives of their own troops. This was done by encouraging local people to bear the burden of the fight. The government of South Vietnam and elements of the population in south Lebanon accepted this commitment. The Vietnamization and what could have been called the "Lebanonization" of the conflict was not easy to accomplish. South Vietnam and the ASL (Army of South Lebanon), the pro-Israeli militia in Lebanon, were not reliable enough. Some of their troops were not too keen on fighting an enemy made up of their own people. There were others who even directly collaborated with the foe. Indeed, many of them did not want to be seen as serving the interests of foreign powers. This was a known reality in these torn countries, used to a foreign presence in the past, such as France in Vietnam and Syria in Lebanon.

From a military point of view, the South Vietnamese military was too weak to stand by itself.[18] So was the ASL. In spite of assistance from the United States in Vietnam and from Israel in Lebanon, those local allies needed their patrons to put their own soldiers on the ground in order to prevent a disaster. In 1975, a few years after the final retreat of the U.S. military, South Vietnam crumbled, while the ASL collapsed in Lebanon in 2000,when Israel had left that country. Those ramifications also bring to mind the desperate flight of those who found themselves in the dangerous situation of being a former ally of the United States/Israel in a country overrun by their foe. The gathering of a Vietnamese crowd near the gates of the U.S. embassy in Saigon, and that of Lebanese near the gates of the Israeli border, symbolized for the United States in Vietnam and for Israel in Lebanon the grim and bitter end of those wars.

The United States in Vietnam strove to minimize the number of casualties among the local population. Strict orders were formulated and distributed to the troops concerning opening fire with artillery or planes.[19] The IDF followed a similar pattern in Lebanon in the 1990s. Those attempts were part of the effort to earn the "hearts and minds" of the local civilians, which also included providing them with medical and economic support, and so forth. Still, it was not enough to convince the entire population to collaborate with the United States in Vietnam or with Israel in Lebanon.

## Public Opinion and Morale

North Vietnam based its strategy on world public opinion and on the internal pressures inside the United States, so the resentment against the war would lead the United States to withdraw.[20] The Hezbollah did the same, hoping the public disputes inside Israel would push the IDF to retreat. In both wars this turned out to be the right approach, due to several factors. Public opinion in the United States, as well as in Israel during the 1990s, was exposed to reports in the media on the progress of the war. Citizens who watched images directly from the battlefield, especially those of wounded soldiers being evacuated by helicopters, were disturbed by such scenes. As the war went on, it seemed to some that repeating those images symbolized the endless, and even meaningless, essence of the fighting. The war in Vietnam therefore caused intense arguments between politicians, journalists and ordinary citizens in the United States, similar to the reaction within Israel regarding the war in Lebanon. Some believed in the need to implement new methods and even escalate the war, while others called to stop the war immediately. Costly battles on the front line intensified public criticism at home. Those disagreements had their effect on the running of the wars in Vietnam and Lebanon, as well as on their respective endings.

U.S. troops in Vietnam, and Israeli soldiers in Lebanon in the 1990s, were more or less aware of the controversy at home. Among the troops (mostly among officers who believed in the necessity of the war), there was resentment against those opposing it, particularly citizens who were safe back home. Those soldiers were angry at civilians who wished to end the war in order to save the lives of the troops, instead of supporting what many of the brass believed to be a just cause. The troops' perspective was that citizens who assumed the war was lost simply did not understand what had actually happened on the battleground, or lacked knowledge, or were misinformed by some in the media who had created a false picture of the outcome of the war.

In spite of the huge geographic gap between the United States and Vietnam, as compared to Israel and Lebanon, American and Israeli troops on the front line had similar ideas and feelings about the inability of their fellow citizens at home to grasp the real situations on the ground. However, some among the American and Israeli troops welcomed, if only secretly, an intervention of civil elements to extract the military from where it was. Those soldiers wished to avoid dangerous (and perhaps useless) presence on enemy territory, hoping they would be spared.

Following the initial evacuation of U.S. troops from Vietnam in 1969, the ones staying behind had serious discipline and morale problems. The soldiers wondered why they had to continue jeopardizing their lives when their military was about to retreat.[21] Israeli troops in Lebanon in the late 1990s had similar feelings after they became aware of the intention to withdraw.

There was also a linkage in this matter between the 2006 war in Lebanon and the conflict in Vietnam. "Widespread Israeli discontent after the 2006 Second Lebanon War provides a contemporary example of popular unrest similar to Americans in the 1970s after the Vietnam Conflict. This portion of U.S. Army history has become a repressed and overlooked memory tucked away in literature and post-war studies."[22]

# 9

# The High- and Low-Intensity War in Libya in 2011

The protests that started in Libya against Muammar al-Gaddafi in mid-February 2011 turned within a few days into an uprising followed by a civil war. Western militaries became involved on 17 April 2011, when the United Nations declared a no-fly zone. In late August of that year, the rebels completed their victory by gaining control of the capital, Tripoli, after having seized most of the state. Gaddafi was found and killed two months later.

The Western powers and the rebels had various strategic and military constraints. The rebels, hoping to depose a dictator who had managed to keep his grip on their country for more than 40 years, had to create their own military strength from scratch. Being civilians and not soldiers, the rebels had to form a militia and learn to operate whatever weapons they could lay their hands on.

Western powers, asked to assist the anti–Gaddafi forces, believed their intervention would prevent a massacre in Libya, as well as a humanitarian disaster that could affect other nearby states, including Europe. Western states wished to get Libyan oil, not waves of African refugees, crossing over the Mediterranean Sea. However, by supporting the rebels and the civil war, Western powers risked more bloodshed and anarchy in Libya, along with huge spending and entanglement in a long conflict.

Western powers had to accomplish their missions without the help of most of the Middle Eastern states, which stayed out of the clash. Israel and Egypt could have contributed to the fight in Libya—Israel through its experience with Arab civil wars, and Egypt as a friendly neighbor on the Libyan border—but neither country did so.

## The Background of Western Intervention

Since the early stages of the conflict, rebels in Libya were not necessarily keen on having Western ground units participate in the skirmish. Air support was another matter. The rebels had to absorb Gaddafi's air bombardments, being unable to foil them, since they lacked planes and a solid anti-aircraft defense. All they could do was look for cover. Gaddafi's air attacks could have delayed and even stopped an advance of the rebels toward Tripoli, since they were completely exposed in the open desert, particularly during the daytime. Western powers, such as the United States, might have tolerated this attack on the rebels for lack of a better option, but air bombardments on unarmed civilians in cities like Benghazi was a different story.

From the start of the civil war in Libya, there was international momentum against Gaddafi, but creating a consolidated coalition was not easy.[1] A no-fly zone, if only over populated areas in east Libya, entailed political and military constraints for NATO.

"The Libyan operation highlighted the importance of obtaining broad political support for operations in the Middle East."[2] There was unwillingness among Western powers such as the United States to invest military resources in the war in Libya. The United States was already carrying a huge burden in Iraq and Afghanistan. Intervention in Libya was mostly a task for the air force and naval aviation, which did not play a key role in the Iraq and Afghanistan theaters, as ground units did. Yet allocating planes and ships jeopardized troops, carrying with it political risks and heavy expenses at a time when there were higher economic and political priorities. The United States could have sent the marines to fight in Tripoli, as they did in the beginning of the 19th century. Gaddafi's center of power was in Tripoli, and the U.S. military has gained vast experience in urban warfare, particularly during the last decade. Yet a campaign in all of Libya might not have been a short one, and after Gaddafi's final defeat the United States could have been entangled in a long struggle to secure Libya, especially if terrorist organizations such as Al-Qaeda had got a strong foothold there.

A lack of clarity about the nature of Gaddafi's opposition seemed to hover over the entire campaign in Libya. The memory of Afghan allies of the United States, who were helped with their mutiny against the Soviet Union and later on turned into sworn enemies of the United States, was not forgotten. Nevertheless, in Western media the Libyan fighters were called *rebels*, which seemed to be a more positive appellation when compared to *insurgents*, a word that had a negative connotation for Western powers because of their fight in Afghanistan and Iraq.

UN resolution 1973 was "to take all necessary measures to protect" the civilians in Libya, "including Benghazi."[3] It followed the advance of Gaddafi's forces toward the gates of Benghazi, which endangered the safety of civilians in several aspects, and primarily those identified as rebels— that is, every civilian who opposed Gaddafi, not necessarily as part of the armed forces opposing the dictator, but also those assisting the rebels one way or another.

Gaddafi's forces were not just a danger to their enemies. Libyan civilians having nothing to do with the uprising were also vulnerable. East Libya, to begin with, was known even before the mutiny as an area where Gaddafi was less popular. Hence, everyone living there was a potential suspect if only because of their presence in cities in east Libya such as Benghazi or Tubruk.

The internal war between the dictator and the rebels would likely have caused the population in east Libya to attempt to escape such a grim fate. This was another negative ramification. Tens (if not hundreds) of thousands of people fleeing from east Libya to the Mediterranean Sea and to the Egyptian border would have created a humanitarian disaster. Of course, not all of the rebels and their supporters in east Libya would have been willing or able to leave. Those staying might have fought to the bitter end, which would have increased the number of civilian casualties, targeted by both sides, deliberately or by mistake, during the battles.

There were no mass massacres of civilians in rebel towns that fell into the hands of Gaddafi's forces. Yet it could have happened in Benghazi, the stronghold of the rebels, as an act of revenge, and in order to crush the uprising once and for all. It was enough for Gaddafi's forces to have besieged Benghazi while bombing it, in order to inflict heavy losses among the people inside the city.

It was difficult for Europe to ignore such a threat against defenseless people on the other side of the Mediterranean Sea, as this location meant negative ramifications for Europe, mostly in terms of immigration and security.

All those possible ominous results caused deep concerns in the West. However, the continuance of the mutiny would have resulted in killing and suffering, particularly among the Libyan population. The rebels might have also clashed with each other—even while fighting against Gaddafi— which would have caused more chaos and casualties. Consequently, it was impossible to know whether the Libyans—like the Iraqis—were not better off tolerating their brutal dictator than facing an internal struggle and uncertainty.

Allowing Gaddafi to crush the uprising could have suppressed the popular protests then taking place around the Arab world. Many in the West considered this wave of change a vital and positive process that would lead to more democratic regimes, which would improve the standard of living among the Arabs and diminish the appeal of joining terrorist groups. Western states hoped that their improved safety would be an indirect (and obviously highly important) result of this trend.

In 2003 the United States and Libya had reached an agreement, but it had problems for both sides.[4] Libya participated in NATO's maneuvers in 2008,[5] and Western powers such as Britain managed to handle Gaddafi, even reaching some level of cooperation with him. However, after he opened fire on his own citizens, particularly in light of what was occurring in the Arab world at that time, it looked like his survival might do more harm than good.

A Gaddafi victory, particularly after the United States and other powers (such as France) openly called for the end of his regime, could have encouraged other regimes in the Middle East to follow his example. Western powers were concerned that regimes such as those in Syria and Iran would copy his methods. However, Assad in Syria acted like Gaddafi anyway, assuming rightly that the probability of Western military involvement against him was quite low.

In 2011 Western powers, including the United States, ignored the aggressive approach of the regime in Bahrain, backed by Saudi Arabia, against the population of Bahrain, since those rulers were considered pro-American. In comparison with Libya, the riots in Bahrain were less bloody, did not last as long and did not evolve into a civil war. Those facts saved Western powers from the need to intervene, as they did in Libya.

## The Western Attack on Libya

An offensive against Libya seemed to Western powers to be the best option, if not the only one, for a quick solution to the crisis. Gaddafi found himself almost isolated in the world once the uprising began, as several steps were taken against him, including sanctions. Nevertheless, Gaddafi was familiar with such a process and assumed it would take a long time to impact him, especially if he did not run out of money to pay his followers and troops.

On the political-military level, the meaning of resolution 1973 was a matter of debate.[6] However, since Gaddafi was a threat to his civilians, Western powers could interpret "all necessary measures" as a green light to neutralize him and resolve the problem. This called for an all-out attack on Gaddafi himself, or at least on his military, particularly the security forces most loyal to Gaddafi, willing to execute every order, including killing and terrorizing the population. The Western powers decided to focus first of all on the Libyan military units that threatened the rebels in Benghazi and Western planes. This meant it was a high-intensity war between an international coalition and the Libyan military.

NATO needed bases near Libya. Italy in particular was essential, because it possessed air and sea ports very close to Libya. There was also the French aircraft carrier *Charles de Gaulle*. Another option was to deploy NATO's forces in the airfields of Libya's neighbors in North Africa, and even in sites that the rebels seized in Libya itself (or those that Western airborne units could seize).

The UN resolution 1973 did not call for an international force under the control of the United Nations,[7] which could have meant creating a vast coalition. Since the war had already started, there was not much time for all of those forces to plan and train together, although NATO, of course, had carried out joint exercises in the past. Even before NATO was established, there had been a history of cooperation between the United States and other Western powers, such as Britain. As in North Africa during the campaigns of World War II, those countries found themselves engaged in a fight against a brutal dictator.

The United States decided not to lead the coalition. Other Western states from NATO also chose not to play a key role, while some of them—such as Germany—stayed completely outside the war. Nevertheless, European powers such as France should have been capable of handling a rather weak rival like Libya. NATO had to intercept the outdated Mig and Sukhoi of Libya, which were not perceived as much of a challenge for the European fighters. France had more than 300 modern jets, such as the Mirage 2000 and the Rafale, which were more than a match for the Libyan air force. The Libyan anti-aircraft batteries might have presented a bigger challenge.

NATO faced military problems in integrated command, intelligence, surveillance and targeting processes, which required U.S. assistance.[8] During the conflict many states in the coalition also suffered from a shortage of ammunition.[9] This was definitely a constraint on operations, and it revealed NATO's state of readiness in high-intensity warfare. In spite of the military potential of Western states, the war in Libya surprised them— obviously not a recommended situation for any military. As a result, Western militaries experienced serious difficulties.

NATO was committed to the fight in Afghanistan, but since that was mostly a ground war NATO's air and sea power could focus on Libya. After the Libyan air force was removed from the sky, the no-fly zone was easier to preserve.

From 31 March to 17 September 2011, NATO launched 8,645 strike sorties, although ammunition was not dropped in all of them.[10] France carried out a third of all the strikes, many of which were launched from the aircraft carrier *Charles de Gaulle*.[11] The Libyan military—not that strong to begin with—was gradually shrinking. Furthermore, its forces were relatively easy to spot from the air, since they were exposed in open terrain, and the front line was quite a small area compared to the entire country.

Gaddafi's forces hid among civilians[12] and found cover in urban areas. Western powers had a dilemma: What would jeopardize the population more—Gaddafi's forces or the air strikes against them? Ironically, using civilians as a human shield made Gaddafi's image even worse, and what was supposed to prevent (or at least reduce) air attacks against him gave Western powers a moral justification to conduct such moves, particularly relying on guided missiles, which diminished the chances of collateral damage. It also compelled Gaddafi's forces to constantly stay very close to civilians, if not dragging them along wherever they went. All in all, there was no way to completely avoid civilian casualties, and the rebels, like free French forces in 1944, accepted air bombardments on their population as part of the price of freedom.

The Libyan military was supposed to possess 500 long-range surface-to-surface missiles. Gaddafi might have assumed that launching them into mainland Europe would, as opposed to stopping the air offensive, probably intensify it, even if the targets of the missiles were not cities but airfields used as a springboard to attack Libya. Gaddafi probably kept the missiles as a last resort, but it was never implemented.

Another last act of defiance and retaliation by Gaddafi, mostly against his own people, could have been destroying the oil infrastructure, like Saddam Hussein did in Kuwait in 1991. Gaddafi's perspective was that if he could not enjoy it, no one else should, particularly his enemies; yet he avoided (or else could not have executed) such retribution.

# The Aims and the Strength of Gaddafi and the Rebels

The civil war in Libya divided the country between the rebels and those who were loyal to Gaddafi and his family. Gaddafi seized control of Libya in 1969 as part of a military coup, at a time when such actions were quite common in the Arab world. In 2011 Libya was swept up in the latest Arab phenomenon: a popular uprising against an autocratic and veteran regime (known in the West as the Arab Spring). Gaddafi claimed he had no official role in Libya. He did not promote himself to the rank of general, either, remaining a colonel, the rank he held when he took power in 1969. Still, he had governed aggressively, enjoying the authority of a president and the status of a king.

During this low-intensity war, each side had its own territory. The rebels were mostly concentrated in east Libya, particularly in the city of Benghazi, while Gaddafi was based in west Libya, in Tripoli (the official capital of the country). Each side had also a grip on other parts of the country, as the rebels had in Misrata. Towns such as Berga changed hands several times during the confrontation. Some of them, like Berga, had strategic value because of their oil infrastructure, which was essential to each side for financing the war while at the same time depriving their adversary of the most vital natural resource in the country. This would have been even more important if the war had lasted for years. Gaddafi, for example, might have had to hire more mercenaries, because he would have not necessarily trusted those from Libya willing to fight for him.

Both sides were not satisfied with the territory they had. In this zero-sum game every side assumed its forces must seize the stronghold of the foe. Aside from the deep hatred between them, each side believed its rival was a clear and present danger.

The war took place between a regime in decline and varied groups that made up the opposition; hence neither side (particularly the rebels) was very organized, possibly because this collision broke out after the uprisings in Tunis and Egypt, which came as a complete surprise.

Before the war Libya's military had 76,000 troops and a few thousand armor vehicles. In February–March 2011 those men and vehicles were split between Gaddafi and the rebels. There were other groups and individuals that served in the military, or else armed themselves from abandoned posts. Some of them probably just wished to protect their lives and their families in that situation of anarchy. In other cases tribes and clans used this opportunity to acquire weapons, while waiting to see which side was winning before openly and fully supporting it. There was also a possibility that both sides, not too strong to begin with, would have worn each other down, thus paving the way for a third party to seize control over the country (or part of it).

## The Ground War

The distance between Benghazi and Tripoli is more than 450 miles, even by the shortest road, which is relatively near the Mediterranean Sea. The pendulum of the campaign went back and forth throughout the war, until the rebels achieved their final victory. During the fighting, when the rebels were on the move, they managed somehow to sustain their advance while seizing towns such as Brega. Tripoli, however, was out of their reach until the end of the war, because of Gaddafi's superior firepower and the rebels' own disorganized force, which delayed them at first, and later on actually stopped them in their tracks.

Gaddafi's forces did not necessarily face an easier task approaching Benghazi. In contrast to the rebels, Gaddafi's forces were better organized, but they had to avoid Western air attacks, and their supply lines became more extended the closer they got to Benghazi. While the rebels traveled light with their civilian cars, assault rifles and rocket-propelled grenades, Gaddafi's forces relied on armor vehicles and artillery, requiring sufficient ammunition, particularly for the long fight on Benghazi, Libya's second largest city.

Logistical difficulties in Libya were compared to those that the British and the Axis Powers experienced in the same area during World War II.[13] In 1941–1943 the two sides were much more professional compared to those fighting in Libya in 2011; yet every time the British or the Axis Powers moved too far away from their respective supply centers, logistical problems began to affect their operations. Although in 2011 the number of fighters participating in the battles was much lower than in the early 1940s, logistical constraints still played a key role. Each side therefore preferred to run operations as close as possible to their main base.

In World War II, the Axis and British troops fighting in Libya sometimes had fuel shortages while maneuvering over what were then undiscovered oil fields. This irony repeated itself in 2011. In spite of the oil facilities in towns along the coast, during the battles, each side had to ensure that its vehicles carried their own fuel, sufficient to advance or retreat as needed.

Gaddafi's forces likely could have mustered enough strength to reach the gates of Benghazi; yet the battle for the city would have been fierce. The rebels were aware of the brutal reprisals their supporters inside the city would have faced had they lost. Gaddafi, however, might have feared that capturing the city could cost him dearly, possibly leaving him too weak to deal with other rebels. The latter included groups that had stayed out of the mutiny up to that point, but who would quickly spot an opportunity to take over at least part of the country. After all, Libya was split between many tribes and clans wishing to expand their influence and gain control of assets such as the oil infrastructure.

In April Gaddafi had about 20,000 men; yet he could not assemble them on the main front line, since they were needed to secure other parts of the country, such as continuing the siege on Misrata. If Gaddafi had risked storming Benghazi with everything he had before the rebels became too strong, the city would have probably been taken. In that case, Western militaries would not have become involved with ground units, as it would have been too late to save the city. As it was, air attacks were essential to stop Gaddafi's forces before they entered the city, while they were in an open territory exposed to air bombardments.

Even if Benghazi had fallen, the rebels would not necessarily have been crushed; they could have continued to fight, particularly in east Libya, as an underground force relying on the friendly population and the vast size of the country. From the beginning they should have perhaps implemented methods of low-intensity war, rather than trying to act as a conventional army confronting Gaddafi directly. However, the entire mutiny was improvised from the start, although it was based on anger that had brewed among many Libyans for decades. The rebels strongly believed they could achieve a quick victory, and they eventually overcame their foe in a matter of months. Considering their starting point, it was quite an accomplishment.

Gaddafi had some reasons to be worried. His regime was almost isolated in the wider world, including the Arab one. He was accused of committing war crimes, and some of his senior officials defected or joined his enemies. Someone from his inner circle might have been encouraged to seize the moment (for example, when Gaddafi's best troops were far away in Benghazi) and remove or kill him. Others in Tripoli, which was full of people who identified with the rebels (as proven when the city ultimately fell), might have had the same idea. Such fears came naturally to Gaddafi, who had always had to be aware of threats against him (mostly internal ones).

The result of the constraints affecting both sides led to a stalemate until the turning point in August. A stalemate was considered by Western powers to be the worst possible outcome, since it meant Gaddafi staying in power. Besides the message it would send to the Arab world (that a dictator could endure if he was willing to fire on his citizens), Gaddafi could have retaliated against Western powers that had turned against him. A deadlock would also have meant high costs of continuing the air campaign over Libya. However, the stalemate helped create a clearer front line, making it easier for Western powers to differentiate between the rebels and Gaddafi's units. It also allowed the rebels much-needed time to organize. The rebels managed to keep the support of their followers—relying on their hatred for Gaddafi—while continuing to rebuild their strength.

## The War at Sea

The war in Libya took place in the air, on the ground and at sea. Both sides had access to other states through the Mediterranean Sea, and they were able to export oil and receive weapons, supplies, and so forth. Most of the important cities and towns in Libya are near the Mediterranean Sea, which made it tempting for each side to conduct sea operations against the other.

From a logistical perspective, the war was relatively easy for NATO to manage, including efforts to prevent the enemy from receiving supplies.[14] Since 23 March 2011, NATO had tried its best to stop weapons, war material and mercenaries from reaching Libya by sea.[15] Libya has six naval bases.[16] The rebels seized ports such as Benghazi and Tubruk, while Gaddafi had Tripoli. Those sites received all kinds of goods and served as springboards for naval operations.

The rebels in the vital city of Misrata were under siege by land; yet they kept in contact with their friends in Benghazi by sea, which allowed them to evacuate wounded and receive food, weapons, and so forth. In 1941, the British in Tubruk had been in a similar situation, although the demands of the rebels in 2011 were more limited, being a smaller and lighter force. The Libyan navy included a few frigates, missile boats and other armed vessels.[17] Warships shelled their enemy and were able to land troops, although a large-scale amphibian operation was probably not an option for either side. Such a move is a complicated challenge even for well-experienced and well-equipped services such as the U.S. Navy and Marines, let alone the Libyan forces. Yet in Libya Gaddafi's forces did not have to face a fierce and organized opposition while creating a bridgehead. It would not have been like forming beachheads in World War II.

## Training the Rebels in Libya

Gaddafi's units had fighter-bombers, gunships, heavy artillery and tanks. The rebels were mostly lightly armed, with weapons that belonged to the Libyan military and had been captured in camps and depots. Consequently, they ran into difficulties matching Gaddafi's forces, particularly in the open desert, where there were not many places to hide from his firepower. In spite of Western air bombardments, Gaddafi's tanks and artillery were often able to push back the rebels. In urban areas, such as key towns near the coast, the rebels sometimes managed to hold on by finding cover in buildings and in the streets. However, as long as the rebels were poorly armed, it was a serious challenge for them to overcome Gaddafi's forces.

The rebels had assault rifles such as the famous AK-47, which was relatively easy to operate, even for such untrained warriors. However, it was effective against troops, not tanks. The rebels could have attacked Gaddafi's armor with their rocket-propelled grenades, but, having a short range (about a thousand yards) and limited penetration capacity, it would have been insufficient. The rebels might have possessed anti-tank missiles with longer range and better penetration, but those missiles were more difficult to handle.

The soldiers who deserted the Libyan military and joined the mutiny were the most professional manpower among the rebels. In spite of the need to organize them into new units, at least in the first months of the war, they were the best (if not the only) men who could have conducted basic combat maneuvers and operate different kinds of weapons.

Western and Arab Special Forces trained the rebels, and arms were sent to them.[18] Britain helped the rebels with both training and advice.[19] Throughout the years Gaddafi had assisted uprisings in several countries in Africa, and it was ironic, if not poetic justice, that he met the same fate when Western states encouraged the mutiny against him.

There were talks between the rebels and Western states taking place in public. There was also a need to hide some aspects of Western assistance and involvement. Although Gaddafi's military was made up of foreign fighters and mercenaries, revealing the Western contribution—be it just instruction and not fighting—could have embarrassed the Western powers. There was international uneasiness regarding the freedom of action Western powers took in implementing military measures against Libya, following UN resolution 1973. It might have looked like a Western attempt to escalate the war—perhaps even the first step of conquering Libya. This would have brought heavy criticism, mostly from the Arab world, which was already voicing suspicion that such was the intention of the West. The instructors therefore had to stay as far as possible from the front line, particularly when it was fluid, and where there was an increased probability of them being revealed or captured by Gaddafi's troops.

The rebels had no modern military and no professional manpower, except for the soldiers who deserted the Libyan military. Yet there were people with technical backgrounds from their work as civilians, who helped the rebels to acquire knowledge about weapons, such as anti-tank missiles and artillery.

The Western intervention reduced Gaddafi's ability to launch air attacks, but creating a no-fly zone in all of Libya for a long time was an economic and political burden for NATO. Since Gaddafi could have periodically tried to send a few gunships in surprise attacks, the rebel forces should have obtained their own anti-aircraft defense.

Until the implementation of UN resolution 1973, states like Russia and China refused to allow a no-fly zone over Libya. Ironically, the rebels were left with mostly outdated Russian and Chinese anti-aircraft guns and shoulder missiles. After the Western intervention the rebels could have received guidance in how to use more advanced anti-aircraft guns and shoulder missiles. This would have improved the ability of the rebels to shoot down planes and gunships flying in low altitude.

Almost a quarter of Libyans have cars—the highest rate in Africa.[20] As far as mobility is concerned, the rebels were to a large extent a "pick-up army," since they maneuvered with pick-up trucks. The rebels moved their artillery by towing them with civilian cars or actually putting the artillery, such as anti-aircraft guns or rockets, on the vans. The civilian cars not only transferred the artillery from one place to another but also served as a fire base in the battlefield itself. Being unarmored, those cars were vulnerable to enemy fire, tanks and artillery, and when driving them the rebels were limited to using paved roads, while original military vehicles such as tanks made it possible for Gaddafi's forces to maneuver in the dunes of the Libyan desert. Bombing and destroying parts of the road was a good means of slowing down the cars, if not blocking them altogether. The various threats against the rebels' cars were one of the major reasons why they needed air cover, or at least a no-fly zone, over their cities and towns, main roads and the front line. The rebels captured some armored vehicles, but teaching the common rebel how to operate and maintain an armored personal carrier could have taken months. Meanwhile, the rebels relied on their civilian cars to bring them to the battlefield, where they dismounted and fought on foot as infantry.

At the beginning of April 2011 there was some improvement in the rebels' performance, but it was not enough to overcome their foe. The rebel forces were based on about a thousand troops who had deserted the Libyan military and a few hundred poorly trained volunteers.[21] Anthony Cordesman estimated that it would take months to "properly train" the rebels. Indeed, the rebels had to be taught basic tactics with the weapons they already possessed. It was easier and faster to do that rather than instruct them on how to use new weapons.

There are relatively simple methods for surviving under enemy fire, including camouflage, digging foxholes, and so forth. The rebels also had to improve their "hit and run" raids, which were most effective at night, when there were more chances to get closer to Gaddafi's forces without being seen. From a short distance even the rebels' rocket-propelled grenades were effective against tanks, particularly the outdated ones.

The rebels could have foiled air attacks in advance by conquering (or at least infiltrating) air bases in order to destroy the planes there. In World War II, during the campaigns in North Africa, the British S.A.S. conducted such raids. The rebels in Libya did not have to be trained commandos. Since the start of the war they had proved their ability to storm military camps, including airfields.

Providing the rebels with a soldier's appearance—boots, uniform, and so on—along with installing elementary military discipline, was essential in organizing the rebels into paramilitary units, no less than teaching tactics and the operation of weapons.

The rebels lacked discipline, and just having courage and high motivation was insufficient. Without instilling a strong willingness to study and drill like regular troops do, there was not much logic in giving ammunition to amateurs who tended to spend precious bullets in random fire in the air, for the sake of vanity, or to celebrate a local victory.

Special Forces directed air support in Libya.[22] Air-ground cooperation has its technical and tactical difficulties, which are not simple to resolve even for professional militaries, let alone military amateurs like the rebels in Libya. If planes received directions at all from the ground, it was not from the rebels, but rather from Western Special Forces; however, this was not a perfect solution, since there were incidents of "friendly fire" when the rebels were attacked from the air by mistake.

There were terrorist organizations such as Al-Qaeda that could have gotten weapons from Libya.[23] The rebels received communication equipment including GPS trackers,[24] which might have been relatively easy to operate, but it created a dependence on their source for maintenance, spare parts, and so forth. Thus, if the rebels had turned against the West, or their advanced military gear had fallen into the hands of Gaddafi or terror organizations such as Al-Qaeda, it would have been more difficult for those enemies to use such complicated devices in the long run.

"The international community should draw lessons from the failure to develop a stronger army and security institutions in Libya after the 2011 internationally assisted ouster of Qaddafi. When the government is controlled by one social faction, it is in effect more like a militia than a national army, whatever its trappings. Outside powers should take care to ensure that any training and support program involves all moderate social actors and otherwise is a force for unity."[25]

## The Resemblance Between the Wars of 1991 and 2011

The war in Libya in 2011 seemed to have a connection to the war in Iraq during the last decade.[26] There is also a certain resemblance between the conflict in Libya and the war against Iraq in 1991.

Following the invasion of Saddam Hussein's Iraq into Kuwait in 1990, the United Nations called for member nations "to use all necessary means to uphold and implement resolution 660," demanding that Iraq leave Kuwait.[27] As a result, a vast coalition that included Western powers and several major Arab states confronted Iraq, driving Hussein's forces out of Kuwait. In 2011, UN resolution 1973 called to implement "all necessary measures" to protect the civilians in Libya, whose personal security was threatened (like the Kuwaitis had been in 1991). For Saddam Hussein in 1991, Kuwait was not an independent state but one of Iraq's provinces. Likewise, Gaddafi saw the area where most of the rebels were—Cyrenaica, in eastern Libya—not as a separate territory, but rather as a part of his state. For the people in eastern Libya (such as those in Benghazi), their fight was to become liberated after having endured a kind of occupation, as the population in Kuwait was in 1991.

The rebels in Libya knew that securing their position required pushing back their enemy. They strove to dethrone Gaddafi (preferably having him end up imprisoned or killed). In Kuwait, it was hoped that Saddam Hussein would meet the same fate. In both cases, the opposition to the tyrant was based on personal hatred and fear.

In 1991 and 2011 the struggle focused not on an Arab state, Iraq/Libya, but rather on its ruler. In 1991 Saddam Hussein enjoyed more public sympathy within the Arab public than Gaddafi did in 2011. Still, many of the Arab people were aware that Saddam Hussein and Gaddafi were brutal, megalomaniac and paranoid dictators. When Saddam Hussein in 1991 and Gaddafi in 2011 went too far, it helped create a general hostile consensus, making it possible for most Arab governments in 1991 to turn against Hussein and for all the Arab states to isolate Gaddafi in 2011.

Western states collaborated with Saddam Hussein in the 1980s, as they did with Gaddafi prior to his fall, but it was done out of pure necessity. Later on in 1991 with Saddam Hussein, and in 2011 with Gaddafi, there was no remorse about the decision to attack these former allies. If Western powers felt any regret, it regarded their cooperation with those shady dictators in the past. Confronting them demonstrated a desire to erase those former connections by destroying them.

The relationships between Saddam Hussein/Gaddafi and Western states were part of an effort by the latter to guarantee oil flows to Western states. Saddam Hussein and Gaddafi became liabilities when they jeopardized this process. Hussein did it when he invaded Kuwait and threatened the oil fields of Saudi Arabia. Gaddafi refused to give up his position, causing a civil war that disrupted the pumping and delivering of oil from Libya.

The war in 1991 was supposed to be the start of "a new order" in the Middle East after the collapse of the Soviet Union. The war in 2011 was intended to mark the beginning of a new era in the Middle East following the decline of autocrat rules in the region.

The wars in 1991 and 2011 did not involve Iran. Nevertheless, the message of 1991, indicating that Western powers and Arab states would not tolerate the occupation of an Arab state, was in a way pointed toward Iran, a regional power in the Persian Gulf that lusted after several small, rich and vulnerable Arab states. In 2011 there was perhaps another indirect message toward Iran, indicating Western powers' determination to strike at Libya's regime for having fired on its unarmed citizens while they demonstrated against their ruler.

The first stage of the offensives in both 1991 and 2011 was an attack from the air and sea with planes and cruise missiles. These campaigns were aimed at the air defense, air force and ground units of the foe. In both wars it was insufficient to gain a victory, and a land offensive was needed to complete the mission.

# Conclusion

Since it was first established, Israel has fought several kinds of wars: high-intensity, low-intensity and hybrid. Israel experienced a high-intensity war in every decade between the 1940s and the 1980s. At the same time Israel also had to deal with low-intensity wars. In 1982 Israel had its first hybrid war, which occurred again in 2006 and 2008–2009 and in 2014.

## Israel's National Strategy and Its High-Intensity Wars

Israel's national security policy, which was designed to handle high-intensity wars, was created after the 1948–1949 war. This policy contains some fundamental principles. The balance of power was overwhelmingly in favor of the Arabs in terms of population size, land and natural resources, and so forth. Losing a high-intensity war would have jeopardized the existence of the state of Israel. Arab states could have destroyed Israel in one successful offensive, whereas they themselves did not face such a danger. Even after achieving victory over the Arabs, Israel could not force them to recognize its right to exist in the Middle East. Israel also developed nuclear weapons, albeit as a last line of defense, upon realizing that in spite of all the achievements of the Israeli military on the conventional battlefield, the loss of a high-intensity war would mean the loss of everything. Egypt also had an unconventional weapon (gas), but, like Israel, it relied only on its conventional power.

The IDF had to pay special attention to several fronts—particularly the Jordanian border, where Israel lacked strategic depth and was exposed to a surprise attack. Israel could not afford a Jordanian military, or any other Arab force, using the West Bank as a base for an attack, considering the proximity of that territory to the vital Tel Aviv area.

The odds of the Arabs beating Israel increased when they joined forces against the Jewish state. The nature of the Arab-Israeli conflict was such that whenever one Arab country got into a high-intensity war with Israel, other Arab states felt compelled to intervene. Yet in the 1956 and 1967 wars, when Israel invaded Egypt, Syria and the Hashemite kingdom did not assist their Arab ally. Arab states could have very quickly joined a fight against Israel, albeit at the expense of effective coordination, as when Iraq and Jordan helped Syria in the 1973 showdown.

In the Yom Kippur War (1973), Arab states united against Israel—temporarily and on a limited scale—for the first time since 1948. During these battles the Arab coalition expanded, but it continued to be based on Syria and Egypt, even though their alliance was quite fragile. Israel managed to toss Syria out of the Golan, but not out of the coalition, where the Syrian military remained alive and kicking. The 1973 showdown demonstrated to Israel the risks of a high-intensity war. Indeed, over the years Israel has had to watch closely any Arab move suggesting an upcoming offensive, particularly as part of a coalition.

Israel also had allies that fought with it against the Arabs. In a high-intensity war this happened only once—in 1956, when France and Britain joined Israel. The latter depended on the two European powers and actually exposed its civilians and military to bombardment, if only at the beginning of the war. Israel trusted its allies to do its work by destroying the Egyptian air force, which they eventually did after two days of combat. During the era of high-intensity wars, Israel also developed its ties with the United States. The latter tolerated Israeli military operations, as long as the campaign did not cause too many problems for the superpower and did not raise the tension with its nemesis, the Soviet Union.

## Israel's Combat Doctrine in a High-Intensity War

Israel's strategic constraints forced it to reach a fast and decisive victory, in order—among other reasons—to minimize its losses. Israel could not compete with the superior number of Arabs, who could afford to take huge casualties. In 1982, as in 1956, Israel was able to focus solely on one front, and the IDF hardly needed internal lines to quickly transport units from one front to the next, as was necessary in the wars of 1948–1949, 1967 and 1973. In 1956 and 1982 Israel's foes absorbed a blow, but they managed to recover, even provoking Israel by deploying a massive force near its border, as Syria did in the Golan.

The IDF had to rely on early warnings of an Arab attack in order to gather its reserve units as soon as possible. If the battles went on for a long time, it could not deploy a large part of its reserve troops on Arab territory for more than a few months. The citizen-soldiers had to return to their civilian jobs, or Israel's economy would have been jeopardized. Those time constraints left the IDF with a narrow window in which to call up its troops, beat the enemy and go back to normal life.

The IDF launched preventive wars in 1956 and 1982 and a preemptive strike in 1967. The Israeli military doctrine was based on the offensive not only strategically but also operationally and tactically. Deep penetrations of armor were a familiar Israeli move in most wars. Conquering areas and destroying Arab forces were two main goals, often achieved together (for example, seizing Sinai in 1956 and 1967 while annihilating Egyptian units stationed there).

Since the 1956 war, the IDF has depended on the armor and the air force. Arab anti-tank and anti-aircraft missiles were a formidable foe mostly in the 1973 showdown; yet they had many shortcomings in comparison to tank/jet performance. This encouraged the IDF to continue relying on its version of air-land offensive. Furthermore, the deployment of Syrian anti-aircraft batteries in the front line in the Golan did not prevent the IAF from conducting strategic bombardments on the Syrian rear in the 1973 showdown (in fact, the location of anti-aircraft batteries actually made the IAF's job easier). Getting back the Golan in the 1973 war was for Syria more than a mere matter of national pride. Possessing this area was essential for its security, because its capital was a few dozen kilometers from the Israeli border. Israel, however, did not need a ground invasion to reach Damascus, since its powerful air force could have inflicted heavy damages there without risking the Israeli land units.

Israel's navy also implemented an aggressive approach that broke the Arab naval siege in the 1973 war. Yet, as in former wars, there were not any massive landing of Israeli troops at beaches, a maneuver that could have had a substantial impact.

## The Strategic Linkage Between High-Intensity, Hybrid and Low-Intensity Wars

The linkage between Israel's low- and high-intensity wars was demonstrated in several ways. Almost all of its low-intensity wars—the border wars that took place until the 1980s—turned to be one of the causes of high-intensity wars (such as the border wars in the 1950s that led to the 1956 war). The reverse could also be true, as when the 1948–1949 high-intensity war led to the border wars of the 1950s. In other cases, high-intensity wars almost happened as a result of a low-intensity war, as in the crisis between Israel and Egypt in 1960. Jordan in particular was at the center of events in the 1950s and 1960s after the kingdom was dragged into low-intensity wars with Israel, mostly because of the Palestinians.

In the 1956 and 1967 high-intensity wars, Jordan and Syria could have joined Egypt against Israel, their common enemy from both border wars and previous high-intensity wars. But in 1956 Syria and the Hashemite kingdom stayed outside the clash, just as they did in the border wars of the 1950s when Egypt was attacked. In 1967, while a high-intensity war raged in the Sinai, Jordan and Syria—to a large extent—confronted Israel only in a low-intensity war at the border. This lack of Arab solidarity in a high-intensity war had devastating consequences because of the scale and goals of the Israeli offensive, as compared with the border wars.

After Israel was forced to return the Sinai and the Gaza Strip to Egypt in 1957, it could at least find some comfort knowing the Gaza Strip had ceased to be a springboard for assaults into the south of Israel in the upcoming years. The high-intensity war in 1956 proved successful in putting an end to Israel's low-intensity war on the Egyptian border. There was, however, no guarantee. The infiltrations that had slowed down before the 1956 war could resume at any time, particularly when Egypt's military buildup, following its arms deal, had given it enough confidence to use the Palestinians to provoke Israel.

## The IDF in Fighting Low- and High-Intensity Wars

Although the IDF wished to concentrate on preparing for a high-intensity war, it has had to participate in low-intensity wars ever since the early 1950s. The IDF could not have ignored assaults against Israeli targets (mostly civilian ones). Therefore, time and resources intended to train the IDF for a high-intensity war were allocated for dealing with the Palestinians in 1987–1993 and in 2000–2005. Through this activity the Palestinians improved the chances that Arab militaries would succeed in a high-intensity war, but the latter did not use this opportunity to attack Israel. Arab militaries calculated that the IDF possessed advanced capabilities that made it superior to them in high-intensity warfare. Furthermore, Syria—despite its desire to retake the Golan—lost its patron following the collapse of the Soviet Union in 1991, and thus its military and political support, which was essential in a high-intensity war against Israel. Iraq, another major Arab adversary of Israel, endured a devastating blow in 1991 that weakened it dramatically, while Egypt continued to choose peace and not war.

All in all, the possibility of another high-intensity war between Israel and Arab states has diminished during recent decades. Hybrid and low-intensity wars have become more common, taking place between Israel and the Hezbollah and/or the Palestinians. Yet the IDF has had to continue practicing for a high-intensity war with Arab states. Neglecting such training could cost Israel far more casualties and damages than in any other war, except an unconventional one.

As far as the IDF was concerned, running low- and high-intensity wars had their similarities. In both cases there was a need for surprise, speed, conducting the same operations—such as vertical flanking—and so forth. However, Israeli troops were exposed to much more danger in a high-intensity war, where the enemy was stronger and the decision more critical. In a low-intensity war the IDF could usually retreat, and the soldiers could live to fight another day.

Until 1956, the IDF general staff trusted the infantry, the dominating corps in the 1948–1949 war. The infantry did well in the border wars of the early and mid–1950s, while the armor had no opportunity to demonstrate its capabilities. Therefore, the achievements of the infantry in both high- and low-intensity wars heavily influenced the buildup of the IDF. Yet, in the decades following the 1956 war, the IDF relied on the armor and the air force in high-intensity wars and, to a large extent, in border wars as well, such as in "the war over water" in the mid-1960s. Usually only a few troops from the air force and the armor participated in the skirmishes, but their lessons were passed to others in their corps.

Low-intensity wars often presented more difficulties than a high-intensity war. Tiny penetrations, for example, were harder to discover in time to intercept them, whereas a massive Arab invasion had many warning signs that enabled the IDF to be prepared for it. Still, considering the ramifications of a full-scale Arab surprise offensive like in 1973, low-intensity wars, at their worst, were preferable to the more dangerous high-intensity war. However, in high-intensity wars the IDF conquered vast areas and annihilated much bigger forces than those manning the garrison posts in the border wars. But accomplishing tasks in high-intensity wars required the IDF to gather far more soldiers than any concentration of troops in the border wars. The scale and nature of many of the missions of high-intensity warfare, in both defense and offense, were such that the IDF often could not have used the low-intensity wars to prepare for them. There was no alternative to having specific, massive and expensive exercises if one hoped to win a high-intensity war. Furthermore, the experience the IDF gained in low-intensity wars was sometimes not only irrelevant to high-intensity wars but also damaging. In many cases, the IDF became too accustomed to procedures of low-intensity warfare. This created an obstacle in high-intensity wars, as could be seen at the beginning of the 1973 showdown, when Israeli armor units in the Golan fought as though it were another "combat day" in the border wars. This problem went both ways, since high-intensity wars could also have a negative impact on conducting low-intensity wars, as in the major raid in 21 March 1968 in Jordan, around the town of Karameh. Some in the IDF expected a swift success similar to the 1967 showdown, but it did not turn out like that.

From the early 1950s up to at least the 1980s, the IAF feared Arab planes would bomb Israeli cities, causing disarray in the assembling of the reserves. Since the 1980s, the same concern has applied to long-range surface-to-surface missiles launched from Arab states—mostly Syria— during a high-intensity war. During the last decade, the Hezbollah could have done the same in a hybrid war by shooting its rockets and missiles.

## Israel's Campaigns in the Gaza Strip in 1956, 1967, 2008–2009 and 2014

Israel has had four major collisions in the Gaza Strip since 1949. Two of them, in 1956 and 1967, were part of high-intensity wars with Egypt, while those in 2008–2009 and 2014 were against a hybrid foe, the Hamas. Egypt in 1956 and 1967 and the Hamas in 2008–2009 and 2014 were Israel's sworn enemies. In spite of this official position, various negotiations—secret or not, with mediators or not—led to all kinds of understandings and ceasefires between Israel and Egypt in 1949–1967 and between Israel and Hamas since 2007. These diplomatic maneuvers postponed major confrontations but did not prevent them.

A few months after the 1956 war, international pressure forced Israel to leave the Gaza Strip, but after the 1967 showdown Israel occupied the entire Gaza Strip for about 27 years. Following the confrontations of

2008–2009 and 2014, Israel immediately abandoned the land it had seized in the Gaza Strip. There was therefore a major difference between the four wars, as far as the Israeli occupation period in the Gaza Strip was concerned. Furthermore, the two Israeli offensives in 1956 and 2008–2009 dramatically reduced assaults against Israeli targets in the area around the Gaza Strip. This outcome persuaded Israel that a full-scale attack provided a better solution than a series of raids in a low-intensity war. Israel was frustrated, however, that those wars (in 1956, 1967 and 2008–2009) failed to topple the regimes controlling the Gaza Strip. Moreover, the Israeli attacks against Nasser in 1956 and 1967, and against the Hamas in 2008–2009, may have rendered them more hostile toward Israel.

The Egyptian military in 1956 and 1967 could have exploited the Gaza Strip as a jumping-off point into Israel, but the Egyptian tanks did not invade then. The Hamas, on the other hand, before and during the confrontations of 2008–2009 and 2014, launched mortars and rockets that caused casualties and damages, harassing and threatening Israeli citizens and soldiers. Israel had defensive options, but traditional reasons, such as the cost of fortifications, caused it to attack in all four wars.

In 1956, 1967, 2008–2009 and 2014, the defender of the Gaza Strip lacked depth and could not have pushed back the Israeli offensive by launching counterattacks. At best, the Arab forces tried to inflict as many casualties as possible upon the IDF and slow down its advance. While in 1956 and 1967 the entire Gaza Strip was seized in one day or in a few days, in 2008–2009 and 2014 most of it was not taken because of political and military reasons. Capturing that hornet's nest could have been worse than continuing to fight against the Hamas from the outside.

In all four wars in the Gaza Strip, the IDF enjoyed valuable advantages, such as air superiority and the ability to concentrate its efforts on one front. Yet in 1956, 2008–2009 and 2014 Israel's citizen-soldiers had only a few days to get organized before the battles started (although they had a few weeks in 1967). There were also problems in coordination between the corps in 1956 and 1967 but less in 2008–2009 and 2014.

## Israel Versus the PLO in 1982 and the Hezbollah in 2006

Israel had two hybrid wars in Lebanon during 1982 and 2006. Strategically, the PLO in 1982 and the Hezbollah in 2006 had no allies fighting alongside them. In 1982 Israel and Syria clashed in Lebanon, and in 2006 this could have repeated itself. Yet, in contrast to 1982, Syria had much less grip on Lebanon and its forces were not deployed there, which reduced the chances of friction between Israel and Syria.

The more the PLO and the Hezbollah turned into conventional formations, the more they became vulnerable because their men and positions were exposed to Israeli bombardments. As a military force, the Hezbollah in 2006 was better prepared than the PLO in 1982 in terms of fortifications, firepower and hybrid capabilities. The Hezbollah and the PLO tended to fight in small groups; yet the former kept its formations, while the army of the PLO in south Lebanon largely disintegrated. In urban warfare men from those two non-state organizations often performed well. The most powerful hybrid foe the IDF ever confronted was encountered in the 2006 war.

In 1982, several Israeli divisions penetrated dozens of kilometers into Lebanon. Israeli troops often preferred to resort to firepower rather than maneuvering on the tactical and even the operational levels. In 2006 the IDF depended strategically on firepower, as opposed to the traditional land offensive. This was, in a way, justified, since before the 1982 war the IDF had about a year to train for a possible war in Lebanon, while in 2006 the war came as a surprise, finding the IDF troops (mostly the ground units) much less ready to attack. In 2006, the IDF possessed more accurate and sophisticated firepower than in 1982, when from the first stage of the war the massive Israeli offensive conquered substantial parts of Lebanon. In 2006 Israel hesitated to commence a major attack, and it eventually seized much less land compared to 1982, thereby exposing its citizens to dozens of rockets fired daily throughout the 34-day campaign. In 1982 the Israeli population was out of range of the PLO's artillery within a few days. In both wars the IDF did not fully exploit the potential of vertical flanking from the sea and air, due to a fear of taking risks that might have cost casualties.

Campaigns against guerrilla and terror forces, including hybrid rivals, exposed IDF weaknesses, providing it with an opportunity to correct its mistakes prior to a more important future showdown. Thus, the failures of the IDF in the 2006 confrontation could have ensured that its units would be ready for a more serious collision with the Hezbollah. Furthermore, the 2006 clash in Lebanon ended without a clear Israeli victory, if any at all. These unwelcome results urged the IDF to aspire for a quick and decisive triumph in its next confrontation on that front. However, such a result could have led to the same outcome of the 1982 war: occupation of large areas in Lebanon—that is, a potential quagmire. The IDF would have faced a low-intensity war, as in the 1980s and 1990s, which cost the lives of hundreds of Israelis, mostly troops.

## Resemblance in Military Strategy and Doctrine Between Israel and Western States

The strategic resemblance between Israel and Western states (mostly the United States and Britain) has been demonstrated in several aspects. Israel and the United States have both experienced success, such as beating a powerful foe in their respective independence wars, but also their share of failure (in matters like forcing regime change). In some cases, Israel drew lessons from other states' misfortune, such as the defeat of France in 1940, which was a kind of reminder regarding what to avoid.

The IDF implemented doctrinal and operational patterns of Western militaries, such as relying on the offensive, launching deep armor penetrations, taking advantage of the terrain, and using air bombardments to block enemy reinforcements. In other cases, the IDF repeated mistakes of Western militaries, like adapting forward defense. The IDF also deliberated on issues that were extremely familiar to Western militaries, such as combined arms, the status of the armor and air force, fighting low-intensity wars, and so forth. In spite of the focus on conventional wars, Israel developed nuclear weapons, to be kept as a last resort, similar to NATO.

## The Vietnam War and the Israeli-Hezbollah War in the 1990s

The Vietnam War and Israel's war in Lebanon in the 1990s were quite different; yet in both conflicts the United States and Israel confronted a rival that was very familiar with the battlefield and enjoyed the support of a foreign power (the Soviet Union in Vietnam and Syria and Iran in Lebanon). It was soon realized that in a low-intensity war it is important to immediately neutralize, by military or political methods, the patron of the guerrilla and terrorist organizations.

The United States in Vietnam and Israel in Lebanon had an overwhelming advantage in airpower that was not completely exploited, mostly due to political considerations. This means it is not just the amount of overall force one possesses or brings to the front that counts but also how much of it is usable. Many in Israel and the United States, mostly in their militaries, were frustrated that, having won the battles, they nevertheless lost the war owing to politics.

The United States and Israel tried to adjust to the conditions on the ground by using helicopters for various missions and implementing guerrilla tactics. They were successful at times, but never enough to be decisive. It is possible that, had the United States and Israel transformed more of their units into experts in handling low-intensity warfare, the outcome would have been different. Yet their efforts were already being conducted at the expense of preparing for a high-intensity war on another front, although this kind of collision never happened in either case.

The local allies of both the United States and Israel were supposed to be the ones specializing in low-intensity warfare, but their chronic weaknesses, including overdependence on their respective patrons, excluded that possibility. This clarified to the United States and Israel that occupying the territory and beating enemy forces were not a substitute for a strong and reliable local military.

The United States and Israel also had another front—at home. It was nonviolent, but the dispute about the very necessity for these wars was fierce, demonstrating the need for an internal consensus for such action. The lack of such consensus was another major reason that finally caused the United States in Vietnam and Israel in Lebanon to withdraw. By doing so, they basically admitted they had given up.

The Vietnam War was part of the long struggle the United States faced with the Soviet Union. Lebanon was also just another chapter in Israel's ongoing conflict with the Arabs. Both wars turned to be unsuccessful confrontations and a setback for both countries. However, Israel continued to survive in the Middle East, and the United States ultimately won the Cold War; from their strategic perspective, the smaller wars, although important, occurred on a secondary front against a relatively weak enemy who did not present a danger to their survival.

The United States did not return to clash with Vietnam, although that bitter memory lingered on for decades, rising after the wars in Iraq and Afghanistan, because of the similarity in running a counterinsurgency campaign in a Third World country far away from home. During the 2006 war in Lebanon, Israel was haunted by the ghost of the battles of the 1990s, which were against the same hybrid foe, and fought on the same front.

# High- and Low-Intensity Warfare in Libya in 2011

Western powers did not forget their civilians who were murdered in terrorist attacks supported or planned by Muammar al-Gaddafi. They did not forgive him, but they were willing to reconcile with him for a time. Nevertheless, in 2011 Gaddafi once again became an enemy of Western powers, who feared he was about to slaughter his own citizens, cause a humanitarian disaster, disrupt the oil market and have an adverse effect on what seemed to be positive developments in the Middle East. However, Western powers hesitated to intervene in the civil war in Libya, mostly due to financial considerations and two other ongoing wars in Muslim countries. Eventually the Western powers became convinced that military action could be carried out quickly and effectively. European powers had the combat capability to protect Libyan civilians, if only through air bombardments against Gaddafi's forces.

The showdown in Libya in 2011 was a combination of a low-intensity war between Gaddafi and the rebels and a high-intensity war between Gaddafi and Western powers. In the fight between Gaddafi and the rebels, the latter were the weaker side, while in the collision between Western powers and Gaddafi, the former leader of Libya was the underdog. Gaddafi lost both of those wars: the high-intensity one (actually an air campaign conducted by Western powers) and the low-intensity war (his land battle against the rebels).

The Western coalition enforced a no-fly zone and attacked ground targets from the air. At sea Western powers also tried to cut off Gaddafi's supply lines. Gaddafi might have been able to survive the air bombardments and bypass the partial naval blockade, but when he was beaten on the ground it was all over for its regime. The bombardments had a major influence, practically and psychologically, and might have been in certain stages a decisive factor, but the rebels had to do the rest on the ground. It was proven that a relative upgrading of the combat performance of the rebels, aided by Western air support, was sufficient to beat Gaddafi's forces.

The land and air activities occurred separately. Western forces and the rebels had the same enemy on the same front, but they did not appoint one supreme commander to take charge of all operations. It was a challenge for Western powers, let alone the rebels, to manage their own members. There were all kinds of talks between Western states, NATO and the rebels that included exchange of views, information, and so on, but not an ordered structure. The actual cooperation on the battlefield was between rebels and Special Forces, essential for specific assignments such as directing air strikes. Compared with joint operations between allies in combat, like the 1991 war against Iraq, there was not much of a coordinated effort in Libya; yet a victory was achieved within a few months, similar to the campaign to overthrow the Taliban in 2001.

NATO was not willing to dispatch a significant force, not even in the air. States like Norway, Denmark and Belgian allocated less than ten planes each, as it was a relatively small air exercise and not a war. States like France and Britain, dominant in the politics of the conflict, also carried much of the military burden. The United States obviously played a substantial part in the military effort as well. The coalition on the whole was somewhat improvised, because the war took its members by surprise, but, in spite of constraints (including a shortage of ammunition), they did possess modern militaries. In contrast, the rebels in Libya were a bunch of civilians, poorly trained (if at all), who joined ex-soldiers from the Libyan military. This ragtag militia was mostly lightly armed and moved around in pick-up trucks and other civilian cars. Charging with those unprotected vans was quite dangerous, particularly against tanks. The rebels' weakness was exploited by Gaddafi's military, which was not a highly professional force to begin with, but more than enough against the rebels (at least in the early stages of the war). Each side advanced and conquered key towns along the coast, losing ground and regaining it, until the rebels managed to hold their initiative and win. Considering the challenge of arming and training the rebels in Libya, perhaps it was too much of a risk for the Western states to intervene and support the rebels at the beginning of the war, as it could have dragged on and even been lost without a reasonably reliable rebel force on the ground.

There was much at stake for both Western powers and the rebels, particularly the latter. There were strategic and military constraints, both pro and con, when it came to confronting Gaddafi. The rebels, if only out of despair, might have done it anyway, but for the Western powers it was a crisis that turned into an opportunity. Those two very different allies took a gamble, which paid off at least as far as getting rid of Gaddafi.

The main goal of this book was to examine several perspectives of Israel's high-intensity, hybrid and low-intensity wars between the years 1948–2014.

At first Israel was mostly threatened by high-intensity wars, particularly against an Arab coalition, as in 1973, which required developing and testing combat patterns. The IDF seemed to be good at winning this type of confrontation, if only because losing such wars would have led to grave consequences. The low-intensity wars, even at their peak—as in the infiltrations in the 1950s, and the suicide bombers in the confrontation of 2000–2005—did not represent an existential danger to the State of Israel. There was also a strategic and operational linkage between high- and low-intensity wars that sometimes served Israel's interests, while in other cases it did not.

In addition, Israel has faced hybrid foes, mostly in Lebanon in 1982 and 2006. The outcome of both of those wars was decided by two basic factors: the qualities of the Hezbollah as a hybrid force that was better than the PLO, and Israel's mistakes and lack of readiness in 2006 as compared to 1982.

In the Gaza Strip Israel ran four large offensives over the years, in 1956, 1967, 2008–2009 and 2014. The terrain of the Gaza Strip, the ineffectiveness of the defenders, and the performance of the IDF gave the latter decisive advantages in those confrontations.

Another perspective of this book focuses on the resemblance between the strategy and combat doctrines of Israel and the Western states, as shown in the Vietnam War and in Lebanon in the 1990s. The war in Libya in 2011 presented another perspective on high- and low-intensity wars, from both Arab and Western points of view.

All in all, Israel, having since 1948 faced diverse strategic and military challenges, has tried to consolidate its approach depending on the situation at hand. The Arab-Israeli conflict has been a costly, demanding and complicated problem. Israel survived the era of high-intensity wars, but only time will tell if it can deal with the ongoing pressure of hybrid/low-intensity wars, not only in the military arena but also in the political, economic and social ones.

# Chapter Notes

## Introduction

1. http://www.dtic.mil/dtic/tr/fulltext/u2/a502164.pdf.
2. http://usacac.army.mil/CAC2/MilitaryReview/Archives/English/MilitaryReview_20131031_art006.pdf.
3. Christopher O. Bowers, "Identifying Emerging Hybrid Adversaries," *Parameters* (Spring 2012), 39.

## Chapter 1

1. David Ben-Gurion, *Uniqueness and Destiny* (Tel Aviv: Ministry of Defense, 1972), 55.
2. Avi Shlaim, *The Iron Wall* (Tel Aviv: Ydiot Ahronot, 2005), 61–71.
3. Ben-Gurion, *Uniqueness and Destiny*; Shimon Golan, *Hot Border, Cold War* (Tel Aviv: Ministry of Defense, 2000), 26–31.
4. Moshe Dayan, *Story of My Life* (Tel Aviv: Edanim, 1976), 685; Avi Kober, *Military Decision in the Arab-Israeli Wars, 1948–1982* (Tel Aviv: Ministry of Defense, 1995), 394–96.
5. IDF Archives, file number 516/506/1988.
6. Edward C. Keefer, ed., *Arab-Israeli Crisis and War, 1973* (Washington, D.C.: Department of State, United States Government Printing Office, 2011), 658.
7. On Egypt, see Shmuel Bar, *The Yom Kippur War in the Eyes of the Arabs* (Tel Aviv: Ministry of Defense, 1986), 69; On Syria, see Moshe Ma'oz, *Syria and Israel: from War to Peace-Making* (Tel Aviv: Maariv Book Guild, 1996), 122–23.
8. Yigal Allon, *Curtain of Sand* (Tel Aviv:

Hakibbutz Hameuchad, 1960), 35–52; *The Diary of David Ben Gurion*, IDF Archives, 23 August 1953. Kober, *Military Decision*, 147–54; Israel Tal, *National Security* (Tel Aviv: Dvir, 1996).
9. Efraim Inbar, "Time Favors Israel," *Middle East Quarterly* (Fall 2013), http://www.meforum.org/3607/time-favors-israel.
10. On Syria as an actor, see Itamar Rabinovich, *The View from Damascus: State, Political Community and Foreign Relations in Twentieth-Century Syria* (Edgware, UK: Vallentine Mitchell, 2011), 378.
11. Gabriel Ben Dor, "Democratization Processes in the Middle East and the Arab World," in Efraim Inbar, ed., *The Arab Spring, Democracy and Security* (New York: Routledge, 2013), 29.
12. Efraim Inbar, "Israel's Interests in Syria," Begin-Sadat Center for Strategic Studies, http://besacenter.org/perspectives-papers/israels-interests-in-syria/.
13. IDF Archives, file number 62/174/1. Ben-Gurion, *Uniqueness and Destiny*, 189.
14. About the essential nature of the Golan in the case of a Syrian ground attack, see, for example, Uri Bar-Joseph, "Israel's Northern Eyes and Shield: The Strategic Value of the Golan Heights Revisited," *Journal of Strategic Studies* 21, no. 3 (September 1998), 46–66.
15. Michael Eisenstadt, *Arming for Peace?* (Washington, D.C.: Washington Institute for Near East Policy, 1992), 84.
16. Ma'oz, *Syria and Israel*; Itamar Rabinovich, *The Brink of Peace* (Tel Aviv: Yediot Aharonot, 1998); See also http://www.nytimes.com/2008/05/22/world/middleeast/22mideast.html?_r=1&.
17. Aryah Shalev, *Israel and Syria: Peace*

*and Security on the Golan* (Jerusalem: Jaffe Center for Strategic Studies, 1994).

18. Eyal Zisser, "An Israeli Watershed: Strike on Syria: Did Israel's Air Strike Change the Balance of Power?" *Middle East Quarterly* (Summer 2008), 57–62; Giora Eiland, "The IDF: Addressing the Failures of the Second Lebanon War," in Mark A. Heller, ed., *The Middle East Strategic Balance, 2007–2008* (Tel Aviv: Institute for National Security Studies, 2008), 35.

19. Naomi Joy Weinberger, *Syrian Intervention in Lebanon* (New York: Oxford University Press, 1986), 271; Shibley Telhami, "America in Arab Eyes," *Survival* (Spring 2007), 112.

20. See http://news.bbc.co.uk/2/hi/middle_east/4484325.stm.

21. Gershon Rivlin, and Elhanan Oren, eds., *The War of Independence: Ben-Gurion's Diary* (Tel Aviv: Ministry of Defense, 1986).

22. IDF Archives, file number 58/776/8.

23. IDF Archives, file number 59/172/100.

24. Andrea Teti, "A Role in Search of a Hero: Construction and the Evolution of Egyptian Foreign Policy, 1952–1967," *Journal of Mediterranean Studies* 14, no. 1/2 (2004), 90.

25. On the 1967 crisis, see IDF Archives, file number 70/117/206; See also Benjamin Miller, "Balance of Power or the State-to-Nation Balance: Explaining Middle East War Propensity," *Security Studies* (October–December 2006), 699.

26. Shimon Golan, *A War on Three Fronts* (Tel Aviv: Ministry of Defense, 2007), 200–224; IDF Archives, file number 69/522/212; IDF Archives, file number 67/901/1.

27. Mohamed Heikal, *The Road to Ramadan* (London: Collins, 1975), 227.

28. Uri Bar-Joseph, *The Watchmen Fell Asleep: The Surprise of Yom Kippur and Its Sources* (Tel Aviv: Zmore Bitan, 2001), 222–24.

29. Benny Morris, *Righteous Victims: A History of the Zionist-Arab Conflict, 1881–2001* (Tel Aviv: Am Oved, 2003), 370.

30. Avi Kober, and Zvi Ofer, eds., *The Iraqi Army in the Yom Kippur War* (Tel Aviv: Ministry of Defense, 1986).

31. On the 1970 crisis in Jordan, see William B. Quandt, *Decade of Decisions* (Tel Aviv: Ministry of Defense, 1980), 123–46; Avraham Sela, "The 1973 Arab War Coalition: Aims, Coherence, and Gain-Distribution," *Israel Affairs* 6, no. 1 (1999), 50.

32. Chaim Herzog, *The Arab-Israeli Wars* (Jerusalem: Edanim, 1983), 247; See also Kober and Ofer, *The Iraqi Army*, 285; Syed

Ali El-Edroos, *The Hashemite Arab Army, 1908–1979* (Amman, Jordan: The Publishing Committee, 1980), 539; Arie Hashavia, *The Yom Kippur War* (Tel Aviv: Zmora, Bitan, Modan, 1974), 197; Moshe Zak, *King Hussein Makes Peace* (Ramat Gan: Bar Ilan University Press, 1996), 137.

33. Patrick Seale, *Asad of Syria* (Tel Aviv: Ministry of Defense, 1993), 198–203.

34. Saad El Shazly, *The Crossing of the Suez* (Tel Aviv: Ministry of Defense, 1987), 26–28; See also Mohamed Abdel Ghani El Gamasy, *The October War* (Cairo: American University in Cairo Press, 1993), 185–86.

35. Shimon Shamir, *Egypt Under Sadat* (Tel Aviv: Dvir, 1978), 96–100; Risa Brooks, "An Autocracy at War: Explaining Egypt's Military Effectiveness, 1967 and 1973," *Security Studies* (July–September 2006), 425.

36. Shlomo Aronson, *Nuclear Weapons in the Middle East* (Jerusalem: Akademon, 1995), 177.

37. Hanoch Bartov, *Daddo* (Tel Aviv: Maariv Book Guild, 1978), Vol. 1, 306.

38. Shlomo Nakdimon, *Low Probability* (Tel Aviv: Yediot Aharonot, 1982), 62.

39. Heikal, *The Road to Ramadan*, 226–27; George W. Gawrych, *The 1973 Arab-Israeli War: The Albatross of Decisive Victory*, Leavenworth Papers, no. 21 (Fort Leavenworth, KS: Combat Studies Institute, U.S. Army Command and General Staff College, 1996), 55–56.

40. Nadav Safran, *Israel: The Embattled Ally* (Tel Aviv: Schocken, 1979), 277.

41. Israeli Government Session, Israeli State Archives, Vol. 60, 8163/11; Elchnan Oren, *The History of Yom Kippur War* (Tel Aviv: Ministry of Defense, 2013), 307–15; Avraham Adan, *On the Banks of the Suez* (Jerusalem: Edanim, 1979), 170–74.

42. Saad El Shazly, *The Arab Military Option* (San Francisco: American Mideast Research, 1986).

43. http://www.biu.ac.il/Besa/MSPS95 He.pdf.

44. Abner Yaniv, *Politics and Strategy in Israel* (Tel Aviv: Sifriat Poalim, 1994), 127; Avraham Tamir, *A Soldier in Search of Peace* (Tel Aviv: Edanim, 1988), 246.

45. Israel Defense Forces, Air Force History Branch, *From the War of Independence to Operation Kadesh* (Tel Aviv: Ministry of Defense, 1990), 121.

46. Michael Cohen, *Fighting World War Three from the Middle East: Allied Contingency Plans, 1945–1955* (Tel Aviv: Ministry of Defense, 1998), 236.

47. IDF Archives, file number 55/488/247.

48. Erskine B. Childers, *The Road to Suez*

(London: Macgibbon and Kee, 1962), 194–214.

49. Motti Golani, *There Will Be War Next Summer* (Tel Aviv: Ministry of Defense, 1997), Vol. 1, 63.

50. Shimon Peres, *David's Sling* (Jerusalem: Weidenfeld and Nicolson, 1970), 169.

51. David Ben-Gurion, *The Sinai Campaign* (Tel Aviv: Am Oved, 1959), 205.

52. Mordechai Bar-On, *Challenge and Quarrel* (Beer Sheva: Ben-Gurion University, 1991), 253–55; Ehud Yonay, *No Margin for Error* (Jerusalem: Ketr, 1995), 127–28.

53. Air Force History Branch, *From the War of Independence*, 154–55.

54. Air Force History Branch, *From the War of Independence*, 168.

55. Neville Brown, *The Future of Air Power* (London: Croom Helm, 1986), 29; IDF Archives, file number 59/172/100.

56. Air Force History Branch, *From the War of Independence*, 259.

57. Ilan Troen, and Moshe Shemesh, eds., *The Suez-Sinai Crisis, 1956: Retrospective and Reappraisal* (London: Frank Cass, 1990), 281–94; George W. A. Gawrych, *Key to the Sinai: The Battles for Abu Ageila in the 1956 and 1967 Arab-Israeli Wars* (Fort Leavenworth, KS: Combat Studies Institute, U.S. Army Command and General Staff College, 1990), 8.

58. Brown, *The Future of Air Power*, 24; Michael Oren, *Six Days of War* (Tel Aviv: Dvir, 2004), 216–17. Andrew McGregor, *A Military History of Modern Egypt* (London: Praeger Security International, 2006), 270.

59. Shlomo Slonim, "Origins of the 1950 Tripartite Declaration on the Middle East," *Middle Eastern Studies* 23, no. 2 (April 1987), 135–49; Herman Finer, *Dulles Over Suez* (Chicago: Quadrangle Books, 1964), 14.

60. Mordechai Bar-On, *The Gates of Gaza* (Tel Aviv: Am Oved, 1992), 186.

61. David Tal, "Symbol Not Substance? Israel's Campaign to Acquire Hawk Missiles, 1960–1962," *International History Review* 22, no. 2 (June 2000), 314; Abraham Ben-Zvi, *John F. Kennedy and the Politics of Arms Sales to Israel* (London: Frank Cass, 2002), 52–55.

62. IDF Archives, file number 69/9/21.

63. http://www.haaretz.com/news/diplomacy-defense/.premium-1.611001.

64. http://www.eucom.mil/mission/the-region/israel. On U.S. commitment to Israel, see also Lin Noueihed, *The Battle for the Arab Spring: Revolution, Counter-Revolution and the Making of a New Era* (New Haven, CT: Yale University Press, 2012), 16.

65. http://www.jinsa.org/files/2013 GeneralsAndAdmiralsTripReport.pdf.

66. Galia Golan, *Soviet Policies in the Middle East from World War Two to Gorbachev* (Cambridge: Cambridge University Press, 1990), 51–52; Bar-On, *The Gates of Gaza*, 315–17.

67. Raymond L. Garthoff, *D'etente and Confrontation: American-Soviet Relations from Nixon to Reagan* (Washington, D.C.: Brookings Institute, 1985), 382.

68. Shireen T. Hunter, *Iran's Foreign Policy in the Post-Soviet Era* (Santa Barbara, CA: ABC-CLIO, 2010), 8.

69. Avi Shlaim, "Israel and the Conflict," in Alex Danchev and Dan Keohane, eds., *International Perspectives on the Gulf Conflict, 1990–91* (London: St. Martin's Press, 1994), 59–79. See also http://users.ox.ac.uk/~ssfc0005/Israel%20and%20the%20Conflict.html.

70. Paul Danahar, *The New Middle East: The World after the Arab Spring* (New York: Bloomsbury, 2013), 428.

71. http://www.brookings.edu/~/media/Research/Files/Reports/2013/12/05%20centcom%20pollack/Centcom%202013%20Proceedings.pdf.

72. http://www.brookings.edu/research/opinions/2013/10/15-american-role-world-ohanlon-shapiro.

73. Abner Cohen, *Israel and the Bomb* (Tel Aviv and Jerusalem: Shocken, 2000), 96–97; Yaniv, *Politics and Strategy*, 150.

74. http://www.washingtonpost.com/blogs/worldviews/wp/2013/12/02/why-is-the-u-s-okay-with-israel-having-nuclear-weapons-but-not-iran/?1.

75. Anthony H. Cordesman, *Weapons of Mass Destruction in the Middle East* (London: Brassey's, 1991), 141.

76. Gawdat Bahgat, "The Proliferation of Weapons of Mass Destruction: Egypt," *Arab Studies Quarterly* (Spring 2007), 2 Bruce D. Porter, *The USSR in Third World Conflicts* (Cambridge: Cambridge University Press, 1984), 80.

77. IDF Archives, file number, 68/668/261; IDF Archives, file number 71/226/15.

78. Zeev Schiff, *Earthquake in October* (Tel Aviv: Zmora, Bitan, Modan, 1974), 231.

79. Arens Moshe, *Broken Covenant* (Tel Aviv: Yedioth Ahronoth, 1995), 174–75.

80. http://www.brookings.edu/research/testimony/2014/07/16-iran-nuclear-deal-israel-sachs.

81. On Israel and Iran, see Kenneth M. Pollack, *Unthinkable: Iran, the Bomb, and American Strategy* (New York: Simon & Schuster, 2013), 394.

82. David Crist, *The Twilight War* (New York: Penguin Press, 2012).

83. http://www.newrepublic.com/article/

115313/amos-yadlin-iran-strike-why-israel-needs-act-soon.

84. On the subject of containment, see Pollack, *Unthinkable.*

85. On Americans in Israel, see http://www.washingtonpost.com/world/middle_east/for-american-israelis-rift-between-obama-and-netanyahu-is-family-affair/2015/04/10/4743c328-d320-11e4-8b1e-274d670aa9c9_story.html.

86. Ehud Eilam, "Israel's Military Options, Challenges and Constraints in an Attack on Iran," *Defense Studies* 13, issue 1 (April 2013), 1–13.

87. On anti–Israeli declarations, see http://jcpa.org/wp-content/uploads/2012/05/IransIntent2012b.pdf.

88. http://www.reuters.com/article/2013/09/17/us-mideast-nuclear-iaea-idUSBRE98G11C20130917.

89. http://www.atlanticcouncil.org/blogs/menasource/getting-to-zero-in-the-middle-east.

90. On giving up the bomb in exchange for peace, see http://www.washingtonpost.com/blogs/worldviews/wp/2013/12/02/why-is-the-u-s-okay-with-israel-having-nuclear-weapons-but-not-iran/?1.

## *Chapter 2*

1. IDF Archives, file number 55/488/261; Avi Kober, *Military Decision in the Arab-Israeli Wars, 1948–1982* (Tel Aviv: Ministry of Defense, 1995), 168–71. Dov Tamari, *The Armed Nation: The Rise and Decline of the Israel Reserve System* (Tel Aviv: Ministry of Defense, 2012), 158.

2. Ariel Levite, *Offense and Defense* (Tel Aviv: Hakibbutz Hameuchad, 1988), 44.

3. Motti Golani, *There Will Be War Next Summer* (Tel Aviv: Ministry of Defense, 1997), Vol. 1, 29–44.

4. IDF Archives, file number 55/566/36.

5. Benny Morris, *Israel's Border Wars, 1949–1956* (Tel Aviv: Am Oved, 1996), 316. David Tal, "Israel's Road to the 1956 War," *International Journal of Middle East Studies* 28 (1996), 70–71; Michael J. Cohen, "Prologue to Suez: Anglo-American Planning for Military Intervention in a Middle East War, 1955–1956," *Journal of Strategic Studies* 26, no. 2 (June 2003), 159. Elli Lieberman, "What Makes Deterrence Work? Lessons from the Egyptian-Israeli Enduring Rivalry," *Security Studies*, no. 4 (Summer 1995), 875. IDF Archives, file number 84/804/13. Moshe Dayan, *Story of My Life* (Tel Aviv: Edanim, 1976), 264–71. Golani, *There Will Be War Next Summer*, Vol. 1, 86–87.

6. Edgar O'Ballance, *The Sinai Campaign* (London: Faber and Faber, 1959), 46–47. Israel Beer, *Israel's Security* (Tel Aviv: Amikam, 1966), 245–46.

7. Mordechai Bar-On, *Challenge and Quarrel* (Beer Sheva: Ben-Gurion University, 1991), 160–64.

8. Morris, *Israel's Border Wars*, 316.

9. Aharon Yariv, *Cautious Assessment* (Tel Aviv: Ministry of Defense, 1998), 175–76.

10. Zeev Schiff and Ehud Ya'ari, *A War of Deception* (Tel Aviv: Schocken, 1984), 101.

11. Uri Bar-Joseph, "The Paradox of Israeli Power," *Survival* (Winter 2004–2005), 150.

12. Uri Sagie, *Lights Within the Fog* (Tel Aviv: Yediot Aharonot Books, 1998), 97–99.

13. Trevor N. Dupuy and Paul Martell, *Flawed Victory* (Fairfax, VA: Hero Books, 1986), 81.

14. Yhospht Harkabi, *War and Strategy* (Tel Aviv: Ministry of Defense, 1990).

15. Avi Shlaim, *The Iron Wall* (Tel Aviv: Ydiot Ahronot, 2005), 403.

16. Israeli Government Session, Israeli State Archives, Vol. 7566/10, 22 March 1960. IDF Archives, file number 62/847/190.

17. Moshe Ma'oz, *Syria and Israel: From War to Peace-Making* (Tel Aviv: Maariv Book Guild, 1996), 232.

18. Patrick Seale, *Asad of Syria* (Tel Aviv: Ministry of Defense, 1993), 266–67, 342–43.

19. Naomi Joy Weinberger, *Syrian Intervention in Lebanon* (New York: Oxford University Press, 1986), 271.

20. Golani, *There Will Be War Next Summer*, Vol. 2, 441–47.

21. Dupuy and Martell, *Flawed Victory*, 81.

22. Gil Merom, *How Democracies Lose Small Wars* (Cambridge: Cambridge University Press, 2003), 161.

23. Avi Shilon, *Begin 1913–1992* (Tel Aviv: Am Oved, 2008), 384.

24. Schiff and Ya'ari, *A War of Deception*, 382.

25. Moshe Dayan, *Yoman Ma'arekhet Sinai* (Tel Aviv: Ham Hasfer, 1965), 82–85.

26. IDF Archives, file number 70/117/206.

27. IDF Archives, file number 55/488/261.

28. IDF Archives, file number 62/174/1. *The Diary of David Ben Gurion*, IDF Archives, 26 August 1953.

29. Yariv, *Cautious Assessment*, 93.

30. IDF Archives, file number 77/717/48. Ami Gluska, *The Israeli Military and the Origins of the 1967 War* (Tel Aviv: Ministry of Defense, 2004), 394. Matitiahu Mayzel,

*The Golan Heights Campaign, June 1967* (Tel Aviv: Ministry of Defense, 2001), 152–53.

31. Israel Tal, *National Security* (Tel Aviv: Dvir, 1996), 166–69.

32. IDF Archives, file number 52/854/96. Levite, *Offense and Defense*, 38. Mohamed Abdel Ghani El Gamasy, *The October War* (Cairo: American University in Cairo Press, 1993), 136.

33. IDF Archives, file number 62/847/77. IDF Archives, file number 63/145/47. Kober, *Military Decision*, 159–61.

34. IDF Archives, file number 58/790/232. IDF Archives, file number 65/1034/937.

35. Yehuda Schiff, ed., *IDF in Its Corps: Army and Security Encyclopedia* (Tel Aviv: Revivim, 1982).

36. IDF Archives, file number 52/1559/178. Michael Carver, "Conventional Warfare in the Nuclear Age," in Peter Paret, ed., *Makers of Modern Strategy* (Princeton, NJ: Princeton University Press, 1986), 798. Michael L. Handel, *Israel's Political-Military Doctrine* (Cambridge, MA: Center for International Affairs, Harvard University, 1973), 68.

37. IDF Archives, file number 53/137/1. Handel, *Israel's Political-Military Doctrine*.

38. Benny Morris, *Righteous Victims: A History of the Zionist-Arab Conflict, 1881–2001* (Tel Aviv: Am Oved, 2003), p. 277.

39. Amiad Brezner, *Wild Broncos: The Development and the Changes of the IDF Armor, 1949–1956* (Tel Aviv: Ministry of Defense, 1999), 415.

40. IDF Archives, file number 77/717/86.

41. Walter Laqueur, *Confrontation: The Middle East and World Politics* (Tel Aviv: Schocken, 1974), 74.

42. Yigal Allon, *Curtain of Sand* (Tel Aviv: Hakibbutz Hameuchad, 1960), 173. See also IDF Archives, file number 58/790/232, and IDF Archives, file number 55/488/247.

43. Bar-On, *Challenge and Quarrel*, 313.

44. George W. A. Gawrych, *Key to the Sinai: The Battles for Abu Ageila in the 1956 and 1967 Arab-Israeli Wars* (Fort Leavenworth, KS: Combat Studies Institute, U.S. Army Command and General Staff College, 1990), 126.

45. Chaim Nadel, *Between the Two Wars* (Tel Aviv: Ministry of Defense, 2006), 74.

46. Uri Bar-Joseph, *The Watchmen Fell Asleep: The Surprise of Yom Kippur and Its Sources* (Tel Aviv: Zmore Bitan, 2001), 185–89.

47. IDF Archives, file number 49/6308/138; Zahava Ostfeld, *Am Army Is Born* (Tel Aviv: Ministry of Defense, 1994), Part I, 197–323.

48. IDF Archives, file number 57/346/65.

49. Ezov Amiram, *Crossing* (Or Yhuda: Zmora, Bitan, Modan, 2011).

50. Dani Asher, *The Syrians on the Borders* (Tel Aviv: Ministry of Defense, 2008), 82–83, 110–12.

51. IDF Archives, file number 66/292/85; Levite, *Offense and Defense*, 38. El Gamasy, *The October War*, 136.

52. IDF Archives, file number 59/161/17. IDF Archives, file number 66/292/85.

53. IDF Archives, file number 84/804/8.

54. IDF Archives, file number 61/291/228.

55. IDF Archives, file number 77/717/86.

56. IDF Archives, file number 58/776/8.

57. IDF Archives, file number 66/292/81.

58. IDF Archives, file number 77/717/86; IDF Archives, file number 77/717/48.

59. Levite, *Offense and Defense*, 67; Clayton R. Newell, *The Framework of Operational Warfare* (New York: Routledge, 1991), 87.

60. Brian Bond, *The Pursuit of Victory* (Oxford: Oxford University Press, 1996), 184.

61. IDF Archives, file number 77/717/86.

62. Chaim Herzog, *The War of Atonement* (Jerusalem: Edanim, 1975), 193.

63. Edward Luttwak, and Dan Horowitz, *The Israeli Army* (New York: Harper and Row, 1975), 129; IDF Archives, file number 58/59/3.

64. IDF Archives, file number 62/847/39.

65. IDF Archives, file number 58/776/7; Bar-On, *Challenge and Quarrel*, 80.

66. IDF Archives, file number 77/717/48; IDF Archives, file number 70/556/2.

67. Dani Asher, *Breaking the Concept* (Tel Aviv: Ministry of Defense, 2003), 199–201.

68. IDF Archives, file number 59/161/17; Avraham Adan, *On the Banks of the Suez* (Jerusalem: Edanim, 1979), 158.

69. Asher, *The Syrians on the Borders*; Levite, *Offense and Defense*, 108–9.

70. The Investigation Committee of the Yom Kippur War, *The Agranet Report*, IDF Archives (1995), 1181.

71. Asher, *The Syrians on the Borders*, 36.

72. Herzog, *The War of Atonement*, 87–112.

73. Investigation Committee, *The Agranet Report*, 1194; Avraham Tamir, *A Soldier in Search of Peace* (Tel Aviv: Edanim, 1988), 245.

74. On Israeli fortifications in the Golan in 2008, see Anthony H. Cordesman, with the assistance of Aram Nerguizian and Lonut C. Popescu, *Israel and Syria: The Military Balance and Prospects of War* (Westport, CT: Praeger Security International, 2008), 223.

75. IDF Archives, file number 55/63/84; IDF Archives, file number 56/7/1.

76. Golani, *There Will Be War Next Summer*, Vol. 1, 209–12.

77. IDF Archives, file number 62/847/36; See also IDF Archives, file number 59/172/100.

78. IDF Archives, file number 83/1210/148.

79. IDF Archives, file number, 83/1210/147.

80. Mustafa Kabahah, *The War of Attrition as Reflected in Egyptian Sources* (Ramat Aviv: Tel Aviv University, 1995), 113.

81. Trevor N. Dupuy, *Elusive Victory* (London: Macdonald and Jane's, 1978), 441; James D. Crabtree, *On Air Defense* (London: Praeger, 1994), 152.

82. Shmuel L. Gordon, *Thirty Hours in October* (Tel Aviv: Maariv Book Guild, 2008); Luttwak and Horowitz, *The Israeli Army*, 347–51.

83. Eliezer Cohen, and Zvi Lavi, *The Sky Is Not the Limit* (Tel Aviv: Maariv Book Guild, 1990), 615; The Syrians tried to explain the defeat; See Mustafa Tlas, ed., *The Israeli Invasion to Lebanon* (Tel Aviv: Ministry of Defense, 1986), 176; Gabriel A. Richard, *Operation Peace for Galilee* (New York: Hill and Wang, 1985), 97–98; Dupuy and Martell, *Flawed Victory*, 141–42.

84. Tamari, *The Armed Nation*, 196; Benjamin Peled, *Days of Reckoning* (Moshav Ben Sheman: Modan, 2005), 343.

85. IDF Archives, file number 273/506/1988; IDF Archives, file number 516/506/1988.

86. On the 1956 war, see IDF Archives, file number 84/804/13; On the 1982 war, see Dupuy and Martell, *Flawed Victory*, 127.

87. Arie Hashavia, *The Yom Kippur War* (Tel Aviv: Zmora, Bitan, Modan, 1974), 197; Herzog, *The War of Atonement*, 229.

88. Azriel Lorber, *Science, Technology and the Battlefield* (Tel Aviv: Kronenberg Professional Books, 1997), 276.

89. Kenneth S. Brower, "Armor in the October War," *Armor* (May–June 1974), 13.

90. Anthony H. Cordesman, *The Arab-Israeli Military Balance and the Art of Operations* (Washington, D.C.: American Enterprise Institute, 1987), 39.

91. Adan, *On the Banks of the Suez*, 71–72.

92. Investigation Committee, *The Agranet Report*, 1475; Emanuel Sakal, "*The Regulars Will Hold!*"? *The Missed Opportunity to Prevail in the Defensive Campaign in Western Sinai in the Yom Kippur War* (Tel Aviv: Maariv Book Guild, 2011), 292–93.

93. IDF Archives, file number 53/137/1.

94. Uri Bialer, *Oil and the Arab-Israeli Conflict, 1948–63* (London: Macmillan Press, 1999), 250.

95. Herzog, *The War of Atonement* (Jerusalem: Edanim, 1975), 233–39.

96. Yehuda Schiff, ed., *IDF in Its Corps: Army and Security Encyclopedia* (Tel Aviv: Revivim, 1982); IDF Archives, file number 61/291/228.

97. Moshe Bar-Kochva, *Chariots of Steel* (Tel Aviv: Ministry of Defense, 1989), 501.

## Chapter 3

1. Benny Morris, *Israel's Border Wars, 1949–1956* (Tel Aviv: Am Oved, 1996), 17.

2. Morris, *Israel's Border Wars*, 18.

3. IDF Archives, file number 52/1559/259.

4. Morris, *Israel's Border Wars*, 24.

5. David Tal, "Israel's Day to Day Security Conception: Its Origin and Development, 1949–1956" (Ben-Gurion University, 1991).

6. Morris, *Israel's Border Wars*, 448.

7. Morris, *Israel's Border Wars*, 211.

8. Tal, "Israel's Day to Day," 102–3.

9. Tal, "Israel's Day to Day," 172–74.

10. IDF Archives, file number 56/8/54.

11. Avi Shlaim, *The Iron Wall* (Tel Aviv: Ydiot Ahronot, 2005), 137.

12. Motti Golani, *There Will Be War Next Summer* (Tel Aviv: Ministry of Defense, 1997), Vol. 2, 27–44.

13. IDF Archives, file number 62/847/31; Mordechai Bar-On, *Challenge and Quarrel* (Beer Sheva: Ben-Gurion University, 1991), 13.

14. Cole C. Kingseed, *Eisenhower and the Suez Crisis of 1956* (Baton Rouge: Louisiana State University Press, 1995), 32.

15. David Tal, "Israel's Road to the 1956 War," *International Journal of Middle East Studies* 28 (1996), 70; Galia Golan, *Soviet Policies in the Middle East from World War Two to Gorbachev* (Cambridge: Cambridge University Press, 1990), 46; Ze'ev Drori, *Israel's Reprisal Policy, 1953–1956: The Dynamics of Military Retaliation* (New York: Frank Cass, 2005), 138.

16. Bar-On, *Challenge and Quarrel*, 43–44; IDF Archives, file number 59/271/116.

17. IDF Archives, file number 62/847/30. Golani, *There Will Be War Next Summer*, Vol. 1, 107.

18. Moshe Ma'oz, *Syria and Israel: From War to Peace-Making* (Tel Aviv: Maariv Book Guild, 1996), 48–58.

19. Jacob Abadi, "Egypt's Policy towards Israel: The Impact of Foreign and Domestic Constraints," *Israel Affairs* 12, no. 1 (2006), 159.

20. Morris, *Israel's Border Wars*, 23; Erskine B. Childers, *The Road to Suez* (London: Macgibbon and Kee, 1962), 97.

21. Stuart A. Cohen, "A Still Stranger Aspect of Suez: British Operational Plans to Attack Israel, 1955–1956," *International History Review* (May 1988), 261–81.

22. Dov Tamari, *The Armed Nation: The Rise and Decline of the Israel Reserve System* (Tel Aviv: Ministry of Defense, 2012), 291–92.

23. IDF Archives, file number 58/776/7.

24. Golani, *There Will Be War Next Summer*, Vol. 1, 354.

25. Morris, *Israel's Border Wars*, 316; Nadav Safran, *Israel: The Embattled Ally* (Tel Aviv: Schocken, 1979), 320.

26. Israel Tal, *National Security* (Tel Aviv: Dvir, 1996), 193; On 1982, see also Uri Sagie, *Lights Within the Fog* (Tel Aviv: Yediot Aharonot Books, 1998), 97–99; Trevor N. Dupuy and Paul Martell, *Flawed Victory* (Fairfax, VA: Hero Books, 1986), 91.

27. Bar-On, *Challenge and Quarrel*, 280; IDF Archives, file number 84/804/13; Moshe Dayan, *Story of My Life* (Tel Aviv: Edanim, 1976), 264–71.

28. Benny Morris, *Righteous Victims: A History of the Zionist-Arab Conflict, 1881–2001* (Tel Aviv: Am Oved, 2003), 285.

29. Morris, *Israel's Border Wars*, 316–17; Shimon Peres, *David's Sling* (Jerusalem: Weidenfeld and Nicolson, 1970), 169.

30. Martin Van Creveld, *The Sword and the Olive: A Critical History of the Israeli Defense Force* (New York: Public Affairs, 1998), 170.

31. IDF Archives, file number 79/1338/40.

32. On number of troops in Yemen, see Moshe Shemesh, *Arab Politics, Palestinian Nationalism and the Six Day War: The Crystallization of Arab Strategy and Nasir's Descent to War, 1957–1967* (Brighton: Sussex Academic Press, 2008), 2.

33. Moshe Gat, "Nasser and the Six Day War, 5 June 1967: A Premeditated Strategy Or An Inexorable Drift to War?" *Israel Affair* 11, no. 4 (October 2005), 631.

34. Barbara W. Tuchman, *Practicing History* (New York: Alfred A. Knopf, 1981), 179.

35. IDF Archives, file number 77/717/86.

36. On the assumption about a possibility of a huge wave of incursions, see IDF Archives, file number 70/117/206.

37. Ma'oz, *Syria and Israel*, 88.

38. Matitiahu Mayzel, *The Golan Heights Campaign, June 1967* (Tel Aviv: Ministry of Defense, 2001), 214.

39. David Kimche, and Dan Bawly, *The Sandstorm* (Tel Aviv: Am Hassefer, 1968), 10.

40. Abraham Ben-Zvi, *The United States and Israel: The Limits of the Special Relationship* (New York: Columbia University Press, 1993), 78.

41. William B. Quandt, *Decade of Decisions* (Tel Aviv: Ministry of Defense, 1980), 55; George W. Ball and Douglas B. Ball, *The Passionate Attachment* (New York: W. W. Norton, 1992), 52.

42. Abraham Zohar, *War of Attrition, 1967–1970* (Israel: ha-Makhon le-Ḥeker Milḥamot Yiśra'el, 2012); Yaacov Bar Siman-Tov, *The Israeli-Egyptian War of Attrition, 1969–1970* (New York: Columbia University Press, 1980).

43. Ze'ev Drori, *Lines of Fire: The War of Attrition on the Israeli Eastern Front, 1967–1970* (Tel Aviv: Ministry of Defense, 2012).

44. Drori, *Lines of Fire*, 23.

45. Tal, *National Security*, 193; Sagie, *Lights Within the Fog*, 97–99; Dupuy and Martell, *Flawed Victory*, 91.

46. Zeev Schiff, and Ehud Ya'ari, *A War of Deception* (Tel Aviv: Schocken, 1984), 384.

47. Schiff and Ya'ari, *A War of Deception*, 123.

48. Schiff and Ya'ari, *A War of Deception*, 163–66.

49. Bar-On, *Challenge and Quarrel*, 252–53; Shlaim, *The Iron Wall*, 178.

50. Uri Bar-Joseph, "The Paradox of Israeli Power," *Survival* (Winter 2004–2005), 150.

51. *The Diary of David Ben Gurion*, IDF Archives, 24 October 1956.

52. Israel Defense Forces, Air Force History Branch, *From the War of Independence to Operation Kadesh* (Tel Aviv: Ministry of Defense, 1990), 139; Bar-On, *Challenge and Quarrel*, 160–64; Dayan, *Story of My Life*, 214–19.

53. Schiff and Ya'ari, *A War of Deception*, 51–65.

54. Schiff and Ya'ari, *A War of Deception*, 348–57.

55. Schiff and Ya'ari, *A War of Deception*, 38.

56. See *Haaretz*, 28 June 2002, B8.

57. Iraq used to give $25,000 to every Palestinian family with a son who became a suicide bomber; See Amos Harel and Avi Isacharoff, *The Seventh War* (Miskal: Yedioth Ahronoth Books and Chemed Books, 2004), 155.

58. Harel and Isacharoff, *The Seventh War*, 65.

59. On smuggling to the Gaza Strip, see Jeff Black and Mel Frykberg, "Egypt Plays Arbitrator," *Middle East* (August–September 2007), 14; Leslie Susser, "Could Hamastan Be

a Step to Peace?" *Jerusalem Report* (23 July 2007), 17.

60. http://www.haaretz.co.il/hasite/spages/1238221.html.

61. Harel and Isacharoff, *The Seventh War*, 156–58, 212.

## *Chapter 4*

1. Daniel Byman, *A High Price: The Triumphs and Failures of Israeli Counterterrorism* (New York: Oxford University Press, 2011), 16.

2. Zahava Ostfeld, *An Army Is Born* (Tel Aviv: Ministry of Defense, 1994).

3. IDF Archives, file number 53/137/1; IDF Archives, file number 55/488/261.

4. IDF Archives, file number 55/488/261.

5. Benny Morris, *Righteous Victims: A History of the Zionist-Arab Conflict, 1881–2001* (Tel Aviv: Am Oved, 2003), 28.

6. Meir Pail, *The Emergence of Zahal (I.D.F)* (Tel Aviv: Zmora, Bitan, Modan, 1979), 143–60.

7. IDF Archives, file number 75/922/550.

8. Mordechai Bar-On, *The Gates of Gaza* (Tel Aviv: Am Oved, 1992), 69.

9. Moshe Dayan, *Yoman Ma'arekhet Sinai* (Tel Aviv: Ham Hasfer, 1965), 52–53; Moshe Dayan, *Story of My Life* (Tel Aviv: Edanim, 1976), 247–48.

10. Dayan, *Story of My Life*, 247; See also Yigal Allon, *Curtain of Sand* (Tel Aviv: Hakibbutz Hameuchad, 1960), 173.

11. IDF Archives, file number 59/172/132.

12. Avi Kober, *Military Decision in the Arab-Israeli Wars, 1948–1982* (Tel Aviv: Ministry of Defense, 1995), 175; Edward Luttwak and Dan Horowitz, *The Israeli Army* (New York: Harper and Row, 1975), 68.

13. Israel Tal, *National Security* (Tel Aviv: Dvir, 1996), 125.

14. Dayan, *Story of My Life*, 112.

15. IDF Archives, file number 62/174/1.

16. Tal, *National Security*, 129; Luttwak and Horowitz, *The Israeli Army*, 113, 117.

17. Luttwak and Horowitz, *The Israeli Army*, 128–30.

18. IDF Archives, file number 53/103/77.

19. IDF Archives, file number 57/627/35.

20. IDF Archives, file number 1955/566/36.

21. Yoav Gelber, *The Emergence of a Jewish Military* (Jerusalem: Yad Izhak Ben-Zvi Institute, 1986), 519–24; IDF Archives, file number 57/627/35.

22. Gelber, *The Emergence of a Jewish Military*, 522.

23. Amiad Brezner, *Wild Broncos: The Development and the Changes of the IDF Armor, 1949–1956* (Tel Aviv: Ministry of Defense, 1999), 386.

24. IDF Archives, file number 84/804/1.

25. Dayan, *Yoman Ma'arekhet Sinai*, 52.

26. IDF Archives, file number 58/776/8; Motti Golani, *There Will Be War Next Summer* (Tel Aviv: Ministry of Defense, 1997), Vol. 2, 493–502.

27. Reuven Gal, *A Portrait of the Israeli Soldier* (New York: Greenwood Press, 1986), 14.

28. Golani, *There Will Be War Next Summer*, Vol. 2, 518.

29. IDF Archives, file number 62/847/36.

30. Israel Defense Forces, Air Force History Branch, *From the War of Independence to Operation Kadesh* (Tel Aviv: Ministry of Defense, 1990), 137–40.

31. IDF Archives, file number 58/776/7.

32. Mordechai Bar-On, *Challenge and Quarrel* (Beer Sheva: Ben-Gurion University, 1991), 268.

33. Golani, *There Will Be War Next Summer*, Vol. 2, 582–84. IDF Archives, file number 59/172/100.

34. IDF Archives, file number 66/292/81; IDF Archives, file number 78/299/419; Jonathan M. House, *Combined Arms Warfare in the Twentieth Century* (Lawrence: University Press of Kansas, 2001), 229; George W. Gawrych, *The Albatross of Decisive Victory* (London: Greenwood Press, 2000), 26.

35. Tal, *National Security*, 138; IDF Archives, file number 70/34/211.

36. IDF Archives, file number 66/517/54.

37. House, *Combined Arms Warfare*, 229.

38. Shmuel L. Gordon, *Thirty Hours in October* (Tel Aviv: Maariv Book Guild, 2008), 29.

39. Eliezer Cohen and Zvi Lavi, *The Sky Is Not the Limit* (Tel Aviv: Maariv Book Guild, 1990), 259–60.

40. IDF Archives, file number 83/1210/147.

41. http://www.fisherlibrary.org.il/Product.asp?ProdID=907.

42. Anthony H. Cordesman, *The Arab-Israeli Military Balance and the Art of Operations* (Washington, D.C.: American Enterprise Institute, 1987), 39.

43. IDF Archives, file number 83/1210/147; Nadav Safran, *Israel: The Embattled Ally* (Tel Aviv: Schocken, 1979), 217.

44. IDF Archives, file number 77/717/86.

45. Gal, *A Portrait of the Israeli Soldier*, 15; IDF Archives, file number 77/717/48.

46. Ze'ev Drori, *Lines of Fire: The War of Attrition on the Israeli Eastern Front, 1967–*

*1970* (Tel Aviv: Ministry of Defense, 2012), 140–44.

47. Benjamin Peled, *Days of Reckoning* (Moshav Ben Sheman: Modan, 2005), 343.

48. Gordon, *Thirty Hours in October*, 88.

49. Peled, *Days of Reckoning*, 417.

50. David Rodman, "Regime-Targeting: A Strategy for Israel," in Efraim Karsh (ed.), *Between War and Peace: Dilemmas of Israeli Security* (London: Frank Cass, 1996), 155–56.

51. Luttwak and Horowitz, *The Israeli Army*, 393.

52. Chaim Nadel, *Between the Two Wars* (Tel Aviv: Ministry of Defense, 2006), 229–41.

53. Emanuel Sakal, *"The Regulars Will Hold!"? The Missed Opportunity to Prevail in the Defensive Campaign in Western Sinai in the Yom Kippur War* (Tel Aviv: Maariv Book Guild, 2011); Anthony H. Cordesman and Abraham R. Wagner, *The Lessons of Modern War* (London: Westview Press, 1990), 39. Tal, *National Security*, 155.

54. IDF Archives, file number 516/506/1988.

55. Sakal, *"The Regulars Will Hold!"?*

56. Abner Yaniv, *Politics and Strategy in Israel* (Tel Aviv: Sifriat Poalim, 1994), 279.

57. Trevor N. Dupuy, and Paul Martell, *Flawed Victory* (Fairfax, VA: Hero Books, 1986), 81.

58. Ibid.

59. About the Egyptian mistake in 1967, see IDF Archives, file number 77/717/86.

60. Zeev Schiff, and Ehud Ya'ari, *Intifada* (Tel Aviv: Schocken, 1990), 167.

61. Amos Gilboa, "The Israel Defense Forces," *Middle East Military Balance* (1992–1993), 165–66.

62. *IAF Magazine*, August 1993, 18.

63. Amos Harel, and Avi Isacharoff, *The Seventh War* (Miskal: Yedioth Ahronoth Books and Chemed Books, 2004), 56.

64. Harel and Isacharoff, *The Seventh War*, 350.

65. See *Haa'artz*, 21 May 2004, B3.

66. IDF Archives, file number 62/122/92.

67. Drori, *Lines of Fire*, 170–71.

68. Zeev Schiff, *Earthquake in October* (Tel Aviv: Zmora, Bitan, Modan, 1974), 142.

69. Zeev Schiff and Ehud Ya'ari, *A War of Deception* (Tel Aviv: Schocken, 1984), 169.

70. Ehud Eilam, *The Next War between Israel and Egypt: Examining a High Intensity War between Two of the Strongest Militaries in the Middle East* (Edgware, UK: Vallentine Mitchell, 2014); See also http://www.amazon.com/The-Next-between-Israel-Egypt/dp/0853038384.

71. Shmuel L. Gordon, *The Second Lebanon War: Strategic Decisions and Their Consequences* (Ben-Shemen: Modan, 2012), 107–10.

72. Dov Tamari, "Could the IDF Be Capable of Changing Following the Second Lebanon War?" *Maarachot* (November 2007), 41.

73. http://www.haaretz.co.il/news/politics/1.1869856.

74. Yuval Steinitz, "Palestinian Guerila Offensive on the Outskirts of Tel Aviv," *Nativ* (November 1998), 69–78.

75. About the "Iron Dome," see http://atwar.blogs.nytimes.com/2012/11/30/success-of-israels-iron-dome-renews-interest-in-missile-defense-systems/ and http://www.biu.ac.il/SOC/besa/perspectives173.html.

## Chapter 5

1. IDF Archives, file number 55/488/247.

2. Cole C. Kingseed, *Eisenhower and the Suez Crisis of 1956* (Baton Rouge: Louisiana University Press, 1995), 131; See also http://www.archives.gov.il/NR/rdonlyres/934A925F-8EBD-4F00–98EB-6F7D40FF689A/0/PagesVol12.pdf.

3. Nadav Safran, *Israel: The Embattled Ally* (Tel Aviv: Schocken, 1979), 374.

4. Bernard Lewis, *The Middle East and the West* (Tel Aviv: Ministry of Defense, 1970), 171; El-Sayed Amin Shalaby, "Egypt's Foreign Policy, 1952–1992," *Security Dialogue* 23, no. 3 (1992), 108; Dov Tamari, *The Armed Nation: The Rise and Decline of the Israel Reserve System* (Tel Aviv: Ministry of Defense, 2012), 283.

5. Michael Oren, *Six Days of War* (Tel Aviv: Dvir, 2004), 311, 345.

6. See http://www.nrg.co.il/online/1/ART1/821/627.html and http://www.liveleak.com/view?i=4d4_1230615118&comments=1.

7. http://www.iba.org.il/bet/?entity=1026745&type=1.

8. Mordechai Bar-On, *The Gates of Gaza* (Tel Aviv: Am Oved, 1992), 141–67.

9. Efraim Halevy, "Israel's Hamas Portfolio," *Israel Journal of Foreign Affairs* 2, no. 3 (2008), 46.

10. Motti Golani, *There Will Be War Next Summer* (Tel Aviv: Ministry of Defense, 1997), Vol. 2, 600.

11. IDF Archives, file number 70/117/206.

12. http://www.mfa.gov.il/MFA/MFAArchive/2000_2009/2006/Gaza+kidnapping+25-Jun-2006.htm.

13. http://www.jinsa.org/fellowship-program/evelyn-gordon/israelis-lose-faith-international-guarantees#.UO2WErbah8Q.

14. Ben Zion Than, "The Sinai Campaign," *Maarachot* (May 1958), 40.

15. Chaim Herzog, *The Arab-Israeli Wars* (Jerusalem: Edanim, 1983), 107.

16. Benny Morris, *Righteous Victims: A History of the Zionist-Arab Conflict, 1881–2001* (Tel Aviv: Am Oved, 2003), 317.

17. http://he.wikipedia.org/wiki/% D7%A8%D7%A6%D7%95%D7%A2%D7%AA _%D7%A2%D7%96%D7%94#.D7.90.D7.95. D7.9B.D7.9C.D7.95.D7.A1.D7.99.D7.99.D7. 94.

18. Mordechai Bar-On, *Challenge and Quarrel* (Beer Sheva: Ben-Gurion University, 1991), 81.

19. Moshe Dayan, *Yoman Ma'arekhet Sinai* (Tel Aviv: Ham Hasfer, 1965), 64.

20. IDF Archives, file number 77/717/48; Ami Gluska, *The Israeli Military and the Origins of the 1967 War* (Tel Aviv: Ministry of Defense, 2004), 394; Edward Luttwak and Dan Horowitz, *The Israeli Army* (New York: Harper and Row, 1975), 370.

21. Scott C. Farquhar, ed., *Back to Basics: A Study of the Second Lebanon War and Operation CAST LEAD* (Fort Leavenworth, KS: Combat Studies Institute Press, U.S. Army Combined Arms Center, 2009), 29.

22. Israel Defense Forces, Air Force History Branch, *From the War of Independence to Operation Kadesh* (Tel Aviv: Ministry of Defense, 1990), 256–57; Than, "The Sinai Campaign," 41.

23. Moshe Dayan, *Story of My Life* (Tel Aviv: Edanim, 1976), 457.

24. Amiad Brezner, *Wild Broncos: The Development and the Changes of the IDF Armor, 1949–1956* (Tel Aviv: Ministry of Defense, 1999), 407.

25. Moshe Bar-Kochva, "Warfare in Urban Area," *Maarachot* (June–July 1990), 8.

26. Shabtai Teveth, *Tanks of Tammuz* (Tel Aviv and Jerusalem: Socken, 1968), 168–75.

27. Beni Michalson, "The Armor in 'Cast Lead,'" *Armor* (in Israel) (January 2009), 11–14.

28. IDF Archives, file number 67/675/101.

29. IDF Archives, file number 70/719/7.

30. *Ma'ariv*, 9 January 2009, 12.

31. On 1956, see Dayan, *Yoman Ma'arekhet Sinai*, 138–39; On 1967, see Shimon Golan, *A War on Three Fronts* (Tel Aviv: Ministry of Defense, 2007), 43, 243.

32. Dayan, *Story of My Life*, 299; Herzog, *The Arab-Israeli Wars*, 106; Than, "The Sinai Campaign," 41.

33. Herzog, *The Arab-Israeli Wars*, 122.

34. Oren, *Six Days of War*, 259–60.

35. Farquhar, *Back to the Basics*, 30, 72; Eran Orteal, "Cast Lead: Lessons for the Operational Concept," *Maarachot* 425 (June 2009), 26.

## Chapter 6

1. See http://www.dtic.mil/dtic/tr/full text/u2/a502164.pdf and http://www. potomacinstitute.org/attachments/120_ Hoffman_JFQ_109.pdf.

2. http://usacac.army.mil/CAC2/MilitaryReview/Archives/English/MilitaryReview_20131031_art006.pdf.

3. Arye Naor, *Cabinet at War* (Tel Aviv: Lahav, 1986), 27–28.

4. Shimon Shiffer, *Snow Ball* (Tel Aviv: Yediot Aharonot Books, 1984), 77–78.

5. Zeev Schiff, and Ehud Ya'ari, *A War of Deception* (Tel Aviv: Schocken, 1984), 91.

6. http://www.newyorker.com/archive/ 2006/08/21/060821fa_fact.

7. http://www.dtic.mil/cgi-bin/GetTRDoc?AD=ADA530150.

8. http://www.inss.org.il/upload/(FILE) 1268645384.pdf.

9. On lack of readiness in 2006, see http://www.haaretz.co.il/hasite/images/ printed/P300108/vino.pdf; See also Amir Rapaport, *Friendly Fire* (Tel Aviv: Maariv Book Guild, 2007), 286.

10. http://carl.army.mil/download/ csipubs/matthewsOP26.pdf.

11. Amos Harel, and Avi Isacharoff, *The Seventh War* (Miskal: Yedioth Ahronoth Books and Chemed Books, 2004).

12. Rapaport, *Friendly Fire*, 286. http:// carl.army.mil/download/csipubs/matthews OP26.pdf.

13. Avi Kober, *Military Decision in the Arab-Israeli Wars, 1948–1982* (Tel Aviv: Ministry of Defense, 1995), 413; Shimshi Elyashiv, *By Virtue of Stratagem* (Tel Aviv: Ministry of Defense, 1995), 221.

14. http://carl.army.mil/download/ csipubs/matthewsOP26.pdf.

15. Schiff and Ya'ari, *A War of Deception*, 29–30.

16. http://www.dtic.mil/cgi-bin/Get TRDoc?AD=ADA530150.

17. Richard A. Gabriel, *Operation Peace for Galilee* (New York: Hill and Wang, 1985), 80, 231.

18. Scott C. Farquhar, ed., *Back to Basics: A Study of the Second Lebanon War and Operation CAST LEAD* (Fort Leavenworth, KS: Combat Studies Institute Press, U.S. Army Combined Arms Center, 2009), 30.

19. James D. Leaf, "Mout and the 1982 Lebanon Campaign: The Israeli Approach," *Armor* (July–August 2000), 9.

20. http://www.washingtoninstitute.org/

uploads/Documents/pubs/PolicyFocus63. pdf.

21. http://www.idf.il/1133–16160-he/ Dover.aspx.

22. Martin Van Creveld, *The Sword and the Olive: A Critical History of the Israeli Defense Force* (New York: Public Affairs, 1998), 296.

23. http://www.potomacinstitute.org/ attachments/120_Hoffman_JFQ_109.pdf. http://www.washingtoninstitute.org/uploads /Documents/pubs/PolicyFocus63.pdf.

24. Schiff and Ya'ari, *A War of Deception*, 171.

25. http://carl.army.mil/download/csi pubs/matthewsOP26.pdf.

26. Van Creveld, *The Sword and the Olive*, 292.

27. http://www.inss.org.il/publications. php?cat=21&incat=&read=84.

28. Mike Eldar, "Flanking from the Sea in the Operation 'Peace for Galilee,'" *Maarachot*, no. 299 (July–August 1985), 27.

29. Van Creveld, *The Sword and the Olive*, 293.

30. Uri Bar-Joseph, "The Paradox of Israeli Power," *Survival* (Winter 2004–2005), 150.

31. Naomi Joy Weinberger, *Syrian Intervention in Lebanon* (New York: Oxford University Press, 1986), 271; Shibley Telhami, "America in Arab Eyes," *Survival* (Spring 2007), 107–22.

32. Schiff and Ya'ari, *A War of Deception*, 166–67; Gabriel, *Operation Peace for Galilee*, 52–53.

33. Avraham Tamir, *A Soldier in Search of Peace* (Tel Aviv: Edanim, 1988), 166.

34. Schiff and Ya'ari, *A War of Deception*, 168, 172.

35. Jonathan D. Zagdanski, "Round 2 in Lebanon: How the IDF Focused on Coin and Lost the Ability to Fight Maneuver War," *Infantry* (September–October 2007), 35.

36. Schiff and Ya'ari, *A War of Deception*, 171

37. Chaim Herzog, *The Arab-Israeli Wars* (Jerusalem: Edanim, 1983), 295; Moshe Palad, "The Military Aspects of the Lebanon War," in Rubik Rozental, ed., *Lebanon: The Other War* (Tel Aviv: Sifriat Poalim, 1983), 116–17.

38. Trevor N. Dupuy, and Martell Paul, *Flawed Victory* (Fairfax, VA: Hero Books, 1986), 89.

39. Gabriel, *Operation Peace for Galilee*, 51; Schiff and Ya'ari, *A War of Deception*, 167.

40. Yehuda Schiff, ed., *The IDF in Its Corps—Army and Security Encyclopedia—Military and Security, part B* (Tel Aviv: Revivim, 1982), 195.

41. Van Creveld, *The Sword and the Olive*, 296.

42. Benny Morris, *Righteous Victims: A History of the Zionist-Arab Conflict, 1881–2001* (Tel Aviv: Am Oved, 2003), 488–89.

43. Abe F. Marrero, "Hezbollah as a Non-State Actor in the Second Lebanon War: An Operational Analysis," in Kendall D. Gott, ed., *Warfare in the Age of Non-State Actors: Implications for the U.S. Army* (Fort Leavenworth, KS: Combat Studies Institute Press, U.S. Army Combined Arms Center, 2007), 292.

44. http://carl.army.mil/download/csi pubs/matthewsOP26.pdf.

45. http://www.dtic.mil/cgi-bin/GetTR-Doc?AD=ADA468848&Location=U2&doc =GetTRDoc.pdf. http://usacac.army.mil/cac2/ cgsc/carl/download/csipubs/WarfareInThe AgeOfNon-StateActors_2007.pdf.

46. http://www.washingtoninstitute.org/ uploads/Documents/pubs/PolicyFocus63. pdf; Zagdanski, "Round 2 in Lebanon," 35; See also http://smallwarsjournal.com/jrnl/ art/the-2006–lebanon-war-a-short-history-part-ii.

47. Schiff and Ya'ari, *A War of Deception*, 168.

48. Ronen Bergman, *Point of No Return* (Or Yehuda: Kinneret, Zmora, Bitan, Dvir, 2007), 183.

49. http://usacac.army.mil/cac2/cgsc/ carl/download/csipubs/WarfareInTheAge OfNon-StateActors_2007.pdf.

50. Dilip Hiro, *Lebanon: Fire and Embers* (London: Weidenfeld and Nicolson, 1993), 95.

51. Schiff and Ya'ari, *A War of Deception*, 38; See also Mordechai Zipori, *In a Straight Line* (Tel Aviv: Yediot Aharonot Books, 1997), 281.

52. Schiff and Ya'ari, *A War of Deception*, 386.

53. http://www.dtic.mil/cgi-bin/Get TRDoc?AD=ADA468848&Location= U2&doc=GetTRDoc.pdf.

54. Uri Sagie, *Lights Within the Fog* (Tel Aviv: Yediot Aharonot Books, 1998), 97–99.

## Chapter 7

1. David Ben-Gurion, *Uniqueness and Destiny* (Tel Aviv: Ministry of Defense, 1972), 43.

2. *The Diary of David Ben Gurion*, IDF Archives, 13 February 1959.

3. Yehoshafat Harkabi, *War and Strategy* (Tel Aviv: Ministry of Defense, 1990), 89.

4. Raymond Cohen, *International Poli-*

*tics: The Rules of the Game* (London: Longman, 1981), 43.

5. Israeli Government Session, Israeli State Archives, Vol. 23, 31 December 1956.

6. See Marc Trachtenberg, "Preventive War and U.S Foreign Policy," *Security Studies* 16, no. 1 (January–March 2007), 1–31.

7. Ben-Gurion, *Uniqueness and Destiny*, 311.

8. B. H. Liddell Hart, *Thoughts on War* (Tel Aviv: Ministry of Defense, 1989), 62.

9. Katriel Ben-Arie, *September 1939* (Tel Aviv: Lavie, 1987), 168–80.

10. IDF Archives, file number 56/636/29.

11. Uri Bar-Joseph, "The Paradox of Israeli Power," *Survival* (Winter 2004–2005), 150.

12. Ezer Weizman, *The Battle for Peace* (Jerusalem: Edanim, 1981), 183.

13. Max Hastings, *The Korean War* (New York: Simon and Schuster, 1987), 116–17.

14. James F. Dunnigan and Austin Bay, *From Shield to Storm* (New York: William Morrow, 1992), 435.

15. David Ben-Gurion, *The Sinai Campaign* (Tel Aviv: Am Oved, 1959), 347.

16. Shimon Peres, *David's Sling* (Jerusalem: Weidenfeld and Nicolson, 1970), 170.

17. Sydney D. Bailey, *Four Arab-Israeli Wars and the Peace Process* (New York: St. Martin's Press, 1990), 121.

18. Ehud Yonay, *No Margin for Error* (Jerusalem: Ketr, 1995), 130–31.

19. *The Diary of David Ben Gurion*, IDF Archives, 23 October 1950.

20. J. F. C. Fuller, *The Conduct of War* (London: Eyre Methuen, 1972), 53.

21. J. Riley, *Napoleon and the World War of 1813: Lessons in Coalition War Fighting* (London: Frank Cass, 2000), 63–201.

22. B. H. Liddell Hart, *The Defence of Britain* (London: Faber and Faber, 1939), 373; B. H. Liddell Hart, *The Tanks* (London: Cassell, 1959), Vol. 1, 457.

23. Georges Lefebvre, *Napoleon* (London: Routledge and Kegan Paul, 1969), 229.

24. Richard A. Preston and Sydney F. Wise, *Men in Arms* (New York: Praeger, 1974), 265.

25. On France and Britain, see Peres, *David's Sling*, 173.

26. Michael Eisenstadt and Kenneth M. Pollack, "Militaries of Snow and Militaries of Sand: The Impact of Soviet Military Doctrine on Arab Militaries," *Middle East Journal* 55, no. 4 (2001), 549–78; See also IDF Archives, file number 62/847/79.

27. F. W. Mellenthin, *Panzer Battles* (Tel Aviv: Ministry of Defense, 1960), 307.

28. Liddell Hart, *The Defence of Britain*, 103.

29. On Britain, see Elizabeth Kier, *Imagining War* (Princeton, NJ: Princeton University Press, 1997), 97; On France, see Stephen Van Evera, *Causes of War* (Ithaca, NY: Cornell University Press, 1999), 232.

30. J. F. C. Fuller, *On Future Warfare* (London: Sifton, Praed & Co., 1928), 323.

31. Azar Gat, "Liddell Hart's Theory of Armored Warfare: Revising the Revisionists," *Journal of Strategic Studies* 19 (March 1996), 25; See also B. H. Liddell Hart, *When Britain Goes to War* (London: Faber and Faber, 1935), 71. B. H. Liddell Hart, *Defence of the West* (London: Cassell, 1950), 269.

32. Martin Van Creveld, *The Sword and the Olive: A Critical History of the Israeli Defense Force* (New York: Public Affairs, 1998), 187.

33. Brian Holden Reid, *Studies in British Military Thought* (Lincoln: University of Nebraska Press, 1998), 24.

34. Mordechai Bar-On, *Challenge and Quarrel* (Beer Sheva: Ben-Gurion University, 1991), 280, 297.

35. Clayton R. Newell, *The Framework of Operational Warfare* (London: Routledge, 1991), 101–3; Julian Thompson, *The Lifeblood of War* (London: Brassey's, 1991), 4–7.

36. Zeev Schiff, *Earthquake in October* (Tel Aviv: Zmora, Bitan, Modan, 1974), 155.

37. Louis C. Peltier, and Etzel G. Pearcy, *Military Geography* (New York: D. Van Nostrand, 1966), 48.

38. Lefebvre, *Napoleon*, 229.

39. Elyashiv Shimshi, *By Virtue of Stratagem* (Tel Aviv: Ministry of Defense, 1995), 204.

40. Schiff, *Earthquake in October*, 136–37; Hanoch Bartov, *Daddo* (Tel Aviv: Maariv Book Guild, 1978), Vol. 2, 167.

41. Peres, *David's Sling*, 166.

42. Benny Morris, *Righteous Victims: A History of the Zionist-Arab Conflict, 1881–2001* (Tel Aviv: Am Oved, 2003), 486–87.

43. Clifford Dowdey, *Lee's Last Campaign* (New York: Barnes and Noble, 1994), 36.

44. Bernard Brody, *War and Politics* (Tel Aviv: Ministry of Defense, 1980), 343.

45. Bartov, *Daddo*, Vol. 1, 291.

46. Ahron Yafh, "Another Passing Thought on the Yom Kippur War," *Maarachot* (September–October 1996), 22.

47. IDF Archives, file number 59/160/49.

48. Schiff, *Earthquake in October*, 107.

49. Moshe Dayan, *Story of My Life* (Tel Aviv: Edanim, 1976), 687.

50. John English, "The Operational Art: Developments in the Theories of War," in B. J. C. McKercher and Michael A. Hennessy (eds.), *The Operational Art* (London: Praeger, 1996), 11.

51. IDF Archives, file number 55/488/23.

52. Ben-Gurion, *Uniqueness and Destiny*, 84.

53. Malcolm Smith, *British Air Strategy between the Wars* (Oxford: Clarendon Press, 1984), 14.

54. Israel Defense Forces, Air Force History Branch, *From the War of Independence to Operation Kadesh* (Tel Aviv: Ministry of Defense, 1990), 45.

55. I. S. O. Playfair, *History of the Second World War: The Mediterranean and Middle East* (London: Her Majesty's Stationery Office, 1960), Vol. 3, 16–17. James Lucas, *Panzer Army Africa* (London: Macdonald and Jane's, 1977), 120; Denis Richards, *Royal Air Force, 1939–1945* (London: Crown Copyright, 1974), 168.

56. Matthew Allen, *Military Helicopter Doctrines of the Major Powers, 1945–1992* (Westport, CT: Greenwood Press, 1993), 22–23, 26.

57. Ariel Levite, *Offense and Defense* (Tel Aviv: Hakibbutz Hameuchad, 1988), 115.

58. William C. Westmoreland, *A Soldier Reports* (Tel Aviv: Ministry of Defense, 1979), 353.

59. V. D. Sokolovskiy, *Soviet Military Strategy* (New York: Crane, Russak, 1975), 252–53.

60. IDF Archives, file number 66/292/85. Brian Bond, *Liddell Hart: A Study of His Military Thought* (London: Cassell, 1977), 258. Edward Luttwak and Dan Horowitz, *The Israeli Army* (New York: Harper and Row, 1975), 213.

61. Dan Schueftan, *Attrition* (Tel Aviv: Ministry of Defense, 1989), 227.

62. Hassan El Badri, Taha El Magdoub and Mohammed dia El Din Zohdy, *The Ramadan War, 1973* (New York: Hippocrene Books, 1974), 83.

63. Arthur E. R. Boak, *A History of Rome to 565 A.D* (New York: Macmillan, 1943), 87.

64. Liddell Hart, *Defence of the West*, 268.

65. *The Diary of David Ben Gurion*, IDF Archives, 13 February 1959.

66. Shelford Bidwell, *Gunners at War* (Tel Aviv: Ministry of Defense, 1974), 158.

67. Douglas Orgill, *The Tank* (Tel Aviv: Ministry of Defense, 1980), 212.

68. Frank Aker, *October 1973* (Hamden, CT: Archon Books, 1985), 41.

69. T. R. Phillips (ed.), *Roots of Strategy* (Harrisburg, PA: Stackpole Books, 1985), 422–23.

70. On the British army, see John Keegan, *The Second World War* (New York: Penguin Books, 1989), 397; On the United States, see Robert R. Palmer, Bell I. Wiley and William R. Keast, *United States Army in World War II: The Army Ground Forces: The Procurement and Training of Ground Combat Troops* (Washington, D.C.: Historical Division, Department of the Army, 1948), 447–48.

71. IDF Archives, file number 55/488/23.

72. IDF Archives, file number 65/1034/383; IDF Archives, file number 9/192/1963; IDF Archives, file number 62/847/173.

73. Dov Tamari, "The Yom Kippur War: Concepts, Estimations and Conclusions," *Maarachot* (October–November 1980), 13.

74. Michael R. Eastman, "American Landpower and the Middle East of 2030," *Parameters* (Autumn 2012), 14.

75. B. H. Liddell Hart, *Deterrent or Defence* (London: Stevens and Sons, 1960), 117.

76. Adam Lowther, "Asymmetric Warfare and Military Thought," in John J. McGrath. ed., *An Army at War: Change in the Midst of Conflict* (Fort Leavenworth, KS: Combat Studies Institute Press, 2005), 112.

77. Daniel Byman, *A High Price: The Triumphs and Failures of Israeli Counterterrorism* (New York: Oxford University Press, 2011), 3.

78. Norman Cigar, "Al Qaida Theater Strategy: Waging a World War," in Norman Cigar and Stephanie E. Kramer, eds., *Al Qaida: After Ten Years of War* (Quantico, VA: Marine Corps University Press, 2011), 45.

79. Michael R. Melillo, "Outfitting Big-War Military with Small War Capabilities," *Parameters* (Autumn 2006), 27.

80. Jubin M. Goodarzi, *Syria and Iran* (New York: Tauris Academic Studies, 2007), 293.

81. Byman, *A High Price*, 14.

82. Mark Lehenbauer, *Orde Wingate and the British Internal Security Strategy during the Arab Rebellion in Palestine, 1936–1939* (Fort Leavenworth, KS: Combat Studies Institute Press, 2012).

83. David A. Charters, and Maurice Tugwell, eds., *Militaries in Low Intensity Conflict* (London: Brassey's Defense Publishers, 1989), 252–53.

84. Christopher O. Bowers, "Identifying Emerging Hybrid Adversaries," *Parameters* (Spring 2012), 47.

85. Jeffrey Sanderson and Jay Miseli, "Six Easy Ways to Lose a War at the Tactical Level," *Armor* (July–August 2007), 7–10.

86. Westmoreland, *A Soldier Reports*, 252, 275.

87. David Maimon, *The Vincible Terror* (Rights reserved to David Maimon, Richon Lzion, 1993).

88. http://usacac.army.mil/CAC2/Military

Review/Archives/English/MilitaryReview_
20131031_art006.pdf.

89. Max Boot, *Invisible Armies* (New
York: Liveright, 2013), 22.

90. Gian Gentile, *Wrong Turn: America's
Deadly Embrace of Counterinsurgency* (London: New Press, 2013), 5.

91. On military education in Afghanistan,
see Jeffrey Michaels, "Laying a Firm Foundation for Withdrawal? Rethinking Approaches
to Afghan Military Education," *Defense Studies* 11, issue 4 (2012), 594–614.

92. On the danger from the Palestinian
security forces, see http://www.israelbehind
thenews.com/library/pdfs/PAforces-HE.pdf.

93. Shabtai Teveth, *The Cursed Blessing*
(Tel Aviv and Jerusalem: Socken, 1969), 28.

94. Yitzhak Rabin, *The War in Lebanon*
(Tel Aviv: Am Oved, 1983), 23.

95. On Afghanistan and the Gaza Strip,
see http://www.securityaffairs.org/issues/
2010/18/kemp.php.

96. John R. McQueney Jr., "MACV's
Dilemma: Changes for the United States and
the Conduct of the War on the Ground in
Vietnam in 1968," in John J. McGrath, ed., *An
Army at War: Change in the Midst of Conflict*
(Fort Leavenworth, KS: Combat Studies
Institute Press, 2005), 248.

97. Kim Coleman, *A History of Chemical
Warfare* (New York: Palgrave Macmillan,
2005).

98. Bruce D. Porter, *The USSR in Third
World Conflicts* (Cambridge: Cambridge
University Press, 1984), 80.

99. Seymour M. Hersh, *The Samson
Option* (Tel Aviv: Yedioth Ahronoth, 1992),
166–67.

## *Chapter 8*

1. John M. Carroll, "America in Vietnam," in John M. Carroll, and Colin F. Baxter,
eds., *The American Military Tradition* (New
York: Rowman and Littlefield, 2007), 265.

2. Andrew Duncan, "Fifty Years On,
Israel Still Tied to Circles of Defence, Part
Two," *Jane's Intelligence Review* (October
1998), 18.

3. Andrew J. R. Mack, "Why Big Nations
Lose Small Wars: The Politics of Asymmetric
Conflict," *World Politics* (January 1975),
178.

4. Jeffrey McCausland, "The Gulf Conflict: A Military Analysis," Adelphi Paper 282
(November 1993), 55.

5. John R. McQueney, Jr., "MACV's
Dilemma: Changes for the United States and
the Conduct of the War on the Ground in
Vietnam in 1968," in John J. McGrath, ed., *An*

*Army at War: Change in the Midst of Conflict*
(Fort Leavenworth, KS: Combat Studies Institute Press, 2005), 251; Michael Maclear,
*10,000 Day War: Vietnam* (Tel Aviv: Ministry of Defense, 1993), 186–88; Henry A.
Kissinger, *American Foreign Policy* (Tel Aviv:
Am Oved, 1974), 85.

6. William C. Westmoreland, *A Soldier
Reports* (Tel Aviv: Ministry of Defense, 1979),
228.

7. http://www.cgsc.edu/carl/download/
csipubs/helmer.pdf.

8. Moshe Tamir, *Undeclared War* (Tel
Aviv: Ministry of Defense, 2006), 20.

9. Ofer Shelah, *The Israeli Army: A Radical Proposal* (Or Yehuda: Zmora, Bitan,
Dvir, 2003), 91.

10. Westmoreland, *A Soldier Reports*, 374.

11. Kenneth Mertel, *The Jumping Mustang* (Tel Aviv: Ministry of Defense, 1971),
32.

12. Mordhi Alon, in *Yedioth Ahronoth*, 19
September 1997, 7.

13. Maclear, *10,000 Day War*, 305.

14. Mordhi Alon, in *Yedioth Ahronoth*, 19
September 1997, 7.

15. Christina Madsen Fishback, "Doctrine
in the Post-Vietnam Era: Crisis of Confidence," in John J. McGrath, ed., *An Army at
War: Change in the Midst of Conflict* (Fort
Leavenworth, KS: Combat Studies Institute
Press, 2005), 624.

16. Dan Syftan, "Beyond the Relative
Advantage," *Maarachot* (March 1998), 72.

17. John A. Nagl, *Counterinsurgency Lessons from Malaya and Vietnam* (London:
Praeger, 2002), 174.

18. James H. Willbanks, *Abandoning Vietnam* (Lawrence: University Press of Kansas,
2004).

19. Westmoreland, *A Soldier Reports*, 258.

20. Kissinger, *American Foreign Policy*,
107; Mack, "Why Big Nations Lose Small
Wars," 177; On unrest in U.S. campuses, see
Richard A. Preston and Sydney F. Wise, *Men
in Arms* (New York: Praeger, 1974), 354.

21. Westmoreland, *A Soldier Reports*, 267.

22. http://usacac.army.mil/CAC2/Military
Review/Archives/English/MilitaryReview_
20131031_art006.pdf.

## *Chapter 9*

1. http://csis.org/publication/seizing-
multilateral-moment-libya.

2. http://nationalinterest.org/commentary/nato-needs-southern-strategy–9769?
page=1.

3. http://www.un.org/News/Press/docs/
2011/sc10200.doc.htm.

4. http://www.washingtoninstitute.org/pubPDFs/PolicyFocus105.pdf.

5. http://www.inss.org.il/upload/(FILE)1275907784.pdf.

6. http://www.strategicstudiesinstitute.army.mil/pubs/display.cfm?pubID=1161.

7. http://www.ajc.org/atf/cf/%7B42d75369-d582–4380–8395-d25925b85eaf%7D/ASSESSING_UNSC_RESOLUTION_ON_LIBYA_032111.PDF.

8. Ivo H. Daalder and James G. Stavridis, "NATO's Victory in Libya: The Right Way to Run an Intervention," *Foreign Affairs* (March/April 2012), 6.

9. http://www.latestbbcnews.com/robert-gates-accuses-nato-allies-advise-an-uncertain-future.html.

10. http://www.globalsecurity.org/jhtml/jframe.html#http://www.globalsecurity.org/military/library/news/2011/09/20110918-oup-update.pdf.

11. http://csis.org/publication/leading-front-europe-and-new-libya.

12. http://csis.org/publication/libyan-uprising-uncertain-trajectory.

13. http://www.reuters.com/article/2011/03/30/us-libya-military-idUSTRE72T5FV20110330.

14. http://www.carlisle.army.mil/USAWC/parameters/Articles/2012spring/Borghard_Pischedda.pdf.

15. http://www.globalsecurity.org/military/world/war/unified-protector.htm.

16. http://www.inss.org.il/upload/(FILE)1275907784.pdf.

17. http://www.inss.org.il/upload/(FILE)1275907784.pdf.

18. http://csis.org/publication/next-steps-libya-egypt-tunisia-and-other-states-new-regimes; http://nationalinterest.org/commentary/nato-needs-southern-strategy-9769?page=1.

19. http://csis.org/publication/leading-front-europe-and-new-libya.

20. http://en.wikipedia.org/wiki/Transport_in_Libya#Highways.

21. http://csis.org/publication/libya-will-farce-stay-us-and-france-and-britain.

22. http://csis.org/publication/libyan-uprising-uncertain-trajectory.

23. http://csis.org/publication/maghrebian-militant-maneuvers-aqim-strategic-challenge.

24. http://www.inss.org.il/upload/(FILE)1275907784.pdf.

25. http://www.clingendael.nl/publication/adversity-and-opportunity.

26. http://csis.org/publication/libya-will-farce-stay-us-and-france-and-britain.

27. http://www.casi.org.uk/info/undocs/gopher/s90/32.

# Bibliography

## Archives

Investigation Committee of the Yom Kippur War, *The Agranet Report*, IDF Archives, 1995.
Israel Defense Forces and Defense Establishment Archives (IDFA).
Israeli State Archives.

## Books and Articles

Adan, A. *On the Banks of the Suez*. Jerusalem: Edanim, 1979.
Allon, Y. *Curtain of Sand*. Tel Aviv: Hakibbutz Hameuchad, 1960.
Asher, D. *Breaking the Concept*. Tel Aviv: Ministry of Defense, 2003.
Bar, S. *The Yom Kippur War in the Eyes of the Arabs*. Tel Aviv: Ministry of Defense, 1986.
Bar-Joseph, U. "Israel's Northern Eyes and Shield: The Strategic Value of the Golan Heights Revisited." *Journal of Strategic Studies* 21, no. 3. September 1998. 46–66.
Bar-Joseph, U. *The Watchmen Fell Asleep: The Surprise of Yom Kippur and Its Sources*. Tel Aviv: Zmore Bitan, 2001.
Bar-Kochva, M. *Chariots of Steel*. Tel Aviv: Ministry of Defense, 1989.
Bar-On, M. *Challenge and Quarrel*. Beer Sheva: Ben-Gurion University, 1991.
Bar-On, M. *The Gates of Gaza*. Tel Aviv: Am Oved, 1992.
Ben Dor, G. "Democratization Processes in the Middle East and the Arab World." In Efrain Inbar. ed. *The Arab Spring: Democracy and Security*. New York: Routledge, 2013. 12–32.
Ben-Gurion, D. *Uniqueness and Destiny*. Tel Aviv: Ministry of Defense, 1972.
Boot, M. *Invisible Armies*. New York: Liveright, 2013.
Bowers, C. O. "Identifying Emerging Hybrid Adversaries." *Parameters*. Spring 2012): 39–50.
Byman, D. *A High Price: The Triumphs and Failures of Israeli Counterterrorism*. New York: Oxford University Press, 2011.
Cfir, A., and Arz, Y., eds. *The IDF in His Core*. Tel Aviv: Rbibim, 1982.
Cigar, N. "Al Qaida Theater Strategy: Waging a World War." In Norman Cigar and Stephanie E. Kramer. eds. *Al Qaida: After Ten Years of War*. Quantico, VA: Marine Corps University Press, 2011. 35–54.
Cohen, E., and Lavi, Z. *The Sky Is Not the Limit*. Tel Aviv: Maariv Book Guild, 1990.
Cohen, M. J. "Prologue to Suez: Anglo-American Planning for Military Intervention in a Middle East War, 1955–1956." *Journal of Strategic Studies* 26, no. 2. June 2003. 152–83.

Cordesman, A. H. *The Military Balance in the Middle East*. London: Praeger, 2004.

Cordesman, A. H., and A. Nerguizian. *The Arab-Israeli Military Balance: Conventional Realities and Asymmetric Challenges*. Washington, D.C.: Center for Strategic and International Studies, 2010.

Cordesman, A. H., with the assistance of A. Nerguizian, and L. C. Popescu. *Israel and Syria: The Military Balance and Prospects of War*. Westport, CT: Praeger Security International, 2008.

Crist, D. *The Twilight War*. New York: Penguin Press, 2012.

Dayan, M. *Story of My Life*. Tel Aviv: Edanim, 1976.

Drori, Z. *Lines of Fire: The War of Attrition on the Israeli Eastern Front, 1967–1970*. Tel Aviv: Ministry of Defense, 2012.

Eastman, M. R. "American Landpower and the Middle East of 2030." *Parameters*. Autumn 2012. 6–17.

Farquhar, S. C. gen. ed. *Back to Basics: A Study of the Second Lebanon War and Operation CAST LEAD*. Fort Leavenworth, KS: Combat Studies Institute Press, US Army Combined Arms Center, 2009.

Gabriel, R. A. *Operation Peace for Galilee*. New York: Hill and Wang, 1985.

Gerges, F. A. "Egypt and the 1948 War: Internal Conflict and Regional Ambition." In A. Shlaim and E. L. Rogan. eds. *The War for Palestine*. Cambridge: Cambridge University Press, 2001. 151–77.

Gilboa, A. "National Security Concept." In Avner Yaniv, Moshe Ma'oz and Avi Kober. eds. *Syria and Israel's National Security*. Tel Aviv: Ministry of Defense 1991. 143–54.

Gluska, A. *The Israeli Military and the Origins of the 1967 War*. Tel Aviv: Ministry of Defense, 2004.

Golan, S. *Hot Border, Cold War*. Tel Aviv: Ministry of Defense, 2000.

Golani, M. *There Will Be War Next Summer*. Tel Aviv: Ministry of Defense, 1997.

Gordon, S. L. *Dimensions of Quality: A New Approach to Net Assessment of Airpower*. Tel Aviv: Jaffee Center for Strategic Studies, 2003.

Gordon, S. L. *The Second Lebanon War: Strategic Decisions and Their Consequences*. Ben-Shemen: Modan, 2012.

Gordon, S. L. *Thirty Hours in October*. Tel Aviv: Maariv Book Guild, 2008.

Gur, M. "Rehabilitation of the Military and Building it for the Future." In A. Cfir and Y. Arz (eds. *The IDF in His Core: Army and Security—Part B*. Tel Aviv: Rbibim, 1982. 143–63.

Haber, E., and Schiff, Z. *Yom Kippur War Lexicon*. Or Yehuda: Kinneret, Zmora, Bitan, Dvir, 2003.

Herzog, C. *The Arab-Israeli Wars*. Jerusalem: Edanim, 1983.

Herzog, C. *The War of Atonement*. Jerusalem: Edanim, 1975.

House, J. M. *Combined Arms Warfare in the Twentieth Century*. Lawrence: University Press of Kansas, 2001.

Icks, R. J. *Famous Tank Battles*. Windsor Berkshire, UK: Profile Publications, 1972.

Inbar, E. "Time Favors Israel." *Middle East Quarterly*. Fall 2013.

Inbar, E., and Sandler, S. "Israel's Deterrence Strategy Revisited." *Security Studies* 3, no. 2. Winter 1993/1994. 330–58.

Israel Defense Forces, Air Force History Branch. *From the War of Independence to Operation Kadesh*. Tel Aviv: Ministry of Defense, 1990.

Itan, Z. "The Syrian Military." In Avner Yaniv, Moshe Ma'oz and Avi Kober. eds. *Syria and Israel's National Security*. Tel Aviv: Ministry of Defense 1991. 155–70.

Jaber, H. *Hezbollah: Born with a Vengeance*. New York: Columbia University Press, 1997.

Jalali, A. A. "Afghanistan in Transition." *Parameters*. Autumn 2010. 1–15.

James, L. "Nasser and His Enemies: Foreign Policy Decision Making in Egypt on the Eve of the Six-Day War." *Middle East Review of International Affairs* 9, no. 2. June 2005. 23–44.

Kam, E. *Surprise Attack*. Tel Aviv: Maarachot, 1990.

Keefer, E. C. gen. ed. *Arab-Israeli Crisis and War, 1973*. Washington, D.C.: Department of State, United States Government Printing Office, 2011.

Keegan, J. *A History of Warfare*. Tel Aviv: Dvir, 1996.

Kipnis, Y. *1973: The Way to War*. Or Yehuda: Kinneret, Zmora, Bitan, Dvir, 2012.

Kober, A. *Military Decision in the Arab-Israeli Wars, 1948–1982*. Tel Aviv: Ministry of Defense, 1995.

Kober, A., and Ofer, Z.. eds. *The Iraqi Army in the Yom Kippur War*. Tel Aviv: Ministry of Defense, 1986.

Laqueur, W. *Confrontation: The Middle East and World Politics*. Tel Aviv: Schocken, 1974.

Lesch, W. D. *The Fall of the House of Assad*. New Haven, CT: Yale University Press, 2012.

Levite, A. *Offense and Defense*. Tel Aviv: Hakibbutz Hameuchad, 1988.

Levitt, M. *Hezbollah: The Global Footprint of Lebanon's Party of God*. Washington, D.C.: Georgetown University Press, 2013.

Liddell Hart, B. H. *History of the Second World War*. London: Book Club Associates, 1970.

Lieberman, E. "What Makes Deterrence Work? Lessons from the Egyptian-Israeli Enduring Rivalry." *Security Studies*, no. 4. Summer 1995. 851–910.

Limbert, W. J. *Negotiating with Iran*. Washington, D.C.: United States Institute of Peace, 2009.

Lorch, N. *History of the Independence War*. Ramt Gan: Masada, 1966.

Luttwak, E., and Horowitz, D. *The Israeli Army*. New York: Harper and Row, 1975.

Macgregor, D. *Transformation Under Fire*. Tel Aviv: Ministry of Defense, 2007.

Macksey, K. *Tank Versus Tank*. Topsfield, MA: Salem House, 1988.

Ma'oz, M. *Syria and Israel: From War to Peace-Making*. Tel Aviv: Maariv Book Guild, 1996.

Mayzel, M. *The Golan Heights Campaign, June 1967*. Tel Aviv: Ministry of Defense, 2001.

McCausland, J. "The Gulf Conflict: A Military Analysis." Adelphi Paper 282. November 1993.

McGregor, A. *A Military History of Modern Egypt*. Westport, CT: Praeger Security International, 2006.

Michaels, J. "Laying a Firm Foundation for Withdrawal? Rethinking Approaches to Afghan Military Education." *Defense Studies* 11, issue 4. 2012. 594–614.

Morris, B. *Israel's Border Wars, 1949–1956*. Tel Aviv: Am Oved, 1996.

Morris, B. *Righteous Victims: A History of the Zionist-Arab Conflict, 1881–2001*. Tel Aviv: Am Oved, 2003.

Nadel, C. *Between the Two Wars*. Tel Aviv: Ministry of Defense, 2006.

Newell, R. *The Framework of Operational Warfare*. London: Routledge, 1991.

O'Ballance, E. *The Sinai Campaign*. London: Faber and Faber, 1959.

Oren, E. *The History of Yom Kippur War*. Tel Aviv: Ministry of Defense, 2013.

Oren, M. *Six Days of War*. Tel Aviv: Dvir, 2004.

Ostfeld, Z. *An Army Is Born*. Tel Aviv: Ministry of Defense, 1994.

Ousby, I. *The Road to Verdun*. New York: Doubleday, 2002.

Pail, M. *The Emergence of Zahal (I.D.F.)*. Tel Aviv: Zmora, Bitan, Modan, 1979.

Pedatzur, R., and Shiek, D. "The Contribution of Marine Power to Israeli Deterrence in the Future Battlefield." *Nativ* 15, no. 3. May 2002. 24–29.

Peled, B. *Days of Reckoning*. Moshav Ben Sheman: Modan, 2005.

Peled, Y. *Soldier*. Tel Aviv: Ma'ariv Book Guild, 1993.

Peres, S. *David's Sling*. Jerusalem: Weidenfeld and Nicolson, 1970.

Playfair, I. S. O. *History of the Second World War: The Mediterranean and Middle East*. London: Her Majesty's Stationery Office, 1960.

Pollack, K. M. *Arabs at War*. Lincoln: University of Nebraska Press, 2004.

Pollack, K. M. *Unthinkable: Iran, the Bomb, and American Strategy*. New York: Simon & Schuster, 2013.

Powell, C. *My American Journey*. New York: Random House, 1995.

Rabin, Y. *Service Notebook*. Tel Aviv: Ma'ariv Book Guild, 1979.

Rabinovich, I. *The Brink of Peace*. Tel Aviv: Yediot Aharonot, 1998.

Rabinovich, I. *The View from Damascus: State, Political Community and Foreign Relations in Twentieth-Century Syria* . Edgware, UK: Vallentine Mitchell, 2011.

Rapaport, A. *Friendly Fire*. Tel Aviv: Maariv Book Guild, 2007.

Richards, D. *Royal Air Force, 1939–1945*. London: Crown Copyright, 1974.

Rodman, D. "'If I Am Not for Myself...' Methods and Motives Behind Israel's Quest for Military Self-Reliance." *Israel Journal of Foreign Affairs* 4, no. 1, 2010. 53–61.

Rothenberg, G. E. *The Anatomy of the Israeli Army*. London: B.T. Batsford, 1979.

Sagie, U. *Lights Within the Fog*. Tel Aviv: Yediot Aharonot Books, 1998.

Sakal, E. *"The Regulars Will Hold!"? The Missed Opportunity to Prevail in the Defensive Campaign in Western Sinai in the Yom Kippur War*. Tel Aviv: Maariv Book Guild, 2011.

Schiff, Z. *Earthquake in October*. Tel Aviv: Zmora, Bitan, Modan, 1974.

Schiff, Z., and Ya'ari, E. *A War of Deception*. Tel Aviv: Schocken, 1984.

Schueftan, D. *Attrition*. Tel Aviv: Ministry of Defense, 1989.

Schwarzkopf, N. H. *It Doesn't Take a Hero*. New York: Linda Grey Bantam Books, 1992.

Segev, S. *Red Sheet*. Tel Aviv: N. Teberski, 1967.

Shazly, S. El. *The Crossing of the Suez*. Tel Aviv: Ministry of Defense, 1987.

Shelah, O. *The Israeli Army: A Radical Proposal*. Or Yehuda: Zmora, Bitan Dvir, 2003.

Shiek, D. "Naval Strategy: Naval Deterrence as Part of the New Strategic Perception." *Nativ* 110, no. 3. May 2006. 27–32.

Shimshi, E. *By Virtue of Stratagem*. Tel Aviv: Ministry of Defense, 1995.

Shimshi, E. *Those Who Bring the Decision*. Tel Aviv: Ministry of Defense, 2007.

Shlaim, A. *The Iron Wall*. Tel Aviv: Ydiot Ahronot, 2005.

Smith, B. "Deter a War or Win it." *Nativ* 2, no. 115. March 2007. 31–37.

Tal, D. "Israel's Day to Day Security Conception: Its Origin and Development, 1949–1956." Ben-Gurion University, 1991.

Tal, D. "Israel's Road to the 1956 War." *International Journal of Middle East Studies* 28. 1996. 59–81.

Tal, I. *National Security*. Tel Aviv: Dvir, 1996.

Tal, N. *Islamic Fundamentalism: The Case of Egypt and Jordan*. Tel Aviv University, 1999.

Tamari, D. *The Armed Nation: The Rise and Decline of the Israel Reserve System*. Tel Aviv: Ministry of Defense, 2012.

Tamir, A. *A Soldier in Search of Peace*. Tel Aviv: Edanim, 1988.

Telem, B. "The Navy in the Yom Kippur War." *Maarachot* 361. November 1998. 58–70.

Thompson, J. *No Picnic*. Tel Aviv: Ministry of Defense, 1992.

Tuchman, B. W. *Practicing History*. New York: Alfred A. Knopf, 1981.

Van Creveld, M. *The Sword and the Olive: A Critical History of the Israeli Defense Force*. New York: Public Affairs, 1998.

Weizman, E. *The Battle for Peace*. Jerusalem: Edanim, 1981.

Weizman, E. *On Eagles' Wings*. Tel Aviv: Maariv Book Guild, 1975.

Yaniv, A. *Politics and Strategy in Israel*. Tel Aviv: Sifriat Poalim, 1994.

Yariv, A. *Cautious Assessment*. Tel Aviv: Ministry of Defense, 1998.

Yonay, E. *No Margin for Error*. Jerusalem: Ketr, 1995.

Yzaki, S. *In the Eyes of the Arabs*. Tel Aviv: Ministry of Defense, 1969.

Zisser, E. *Commanding Syria: Bashar al–Asad and the First Years in Power*. London: I. B. Tauris, 2007.

Zohar, A. *War of Attrition, 1967–1970*. Israel: ha–Makhon le-Ḥeḳer Milḥamot Yiśra'el, 2012.

# Index